FUNDAMENTALS OF STEAM GENERATION CHEMISTRY

FUNDAMENTALS OF STEAM GENERATION CHEMISTRY

Brad Buecker

Copyright ©2000

PennWell Corporation
1421 South Sheridan Road
PO Box 1260
Tulsa, Oklahoma 74101

Cover and book design by John Potter

[CIP]

All rights reserved. No part of this book may be reproduced, stored in a retrieval system, or transcribed in any form or by any means, electronic or mechanical including photocopying or recording, without the prior permission of the publisher.

Printed in the United States of America.

DEDICATION

This book is dedicated to God for giving me what writing talents I may have, and to the memory of my grandparents, Julia and William Spence, and my uncle, Paul Spence, for all the encouragement and love they gave me throughout their lives.

ACKNOWLEDGMENTS

I would like to thank my friends at BetzDearborn, Drew Industrial, U.S. Filter, Osmonics, Purolite, Jonas & Consultants, Sentry Equipment Corporation, and Marley Cooling Tower for once again providing information that I have used in a technical document. Particular individuals from these organizations include Ray Post, Doug DeWitt-Dick, Steve Dominick, Don Walter, Terry Heller, Lee Machemer, James Groose, and Terry Dwyer, respectively. Special thanks to BetzDearborn for allowing me to use so many of the wonderful illustrations from their water chemistry handbook. Thanks also to the American Society of Mechanical Engineers for the boiler and feedwater chemistry data listed in Chapter 5. I would also like to acknowledge the Electric Power Research Institute. At many of the water treatment and power conferences I have attended in the last decade, one or more speakers will show slides that contain data or diagrams originally produced by EPRI. Because the information in these presentations is in the public domain, references are often not provided. I used some of this public information in this book, but in all cases that I could think of, I tried to acknowledge the original source.

For those readers interested in attending water treatment and/or power conferences, some of the best include the annual International Water Conference held in Pittsburgh, Pennsylvania, the annual Electric Utility Chemistry Workshop in Champaign, Illinois, and the annual Power-Gen conference hosted by PennWell Publishing in various locations. If you need to learn more about steam plant chemistry or operation, I would recommend all three.

I wish to once again thank my friends and former colleagues at City Water, Light & Power, Springfield, Illinois. I learned a lot of chemistry during my twelve years at this facility, in part because I was allowed to work on many different projects. I will always be grateful for the trust you showed in me and for your continued friendship.

Contents

Introduction xiii

1 General Chemistry 1

2 The Chemistry of Natural Water Supplies 35

3 Makeup Treatment 59

4 Fundamentals of Steam Generation and
   Sources of Contamination 133

5 Chemical Treatment Programs for
   Steam Generating Systems................... 185

6 Monitoring Techniques & Control Guidelines....... 225

7 Cooling Water Chemistry..................... 265

Bibliography 315

Index 319

FIGURES

1–1	Periodic Table	7
1–2	Electronegativities	18
1–3	An Electrochemical Reaction	28
2–1	Carbon Dioxide Molecule	36
2–2	Water Molecule	37
2–3	Relationship of CO_2, HCO_3^-, and CO_3^{-2} in water	43
2–4	Solubility of Gypsum and Calcium Carbonate vs. Temperature	54
3–1	Outline of a Common Surface Water Pretreatment Scheme	60
3–2	Outline of Mixing and Settling Zones in a Clarifier	63
3–3	Amine General Structure	64
3–4	Non-ionic Functional Group	65
3–5	An-ionic Functional Group	65
3–6	Clarifier Arrangement I	66
3–7	Clarifier Arrangement II	66
3–8	Clarifier Arrangement III	67
3–9	Calcium Removal vs. Carbonate Alkalinity	74
3–10	Magnesium Removal vs. pH Source	75
3–11	Silica Reduction Possible in the Softening Process	76
3–12	Example of a Groundwater Pretreatment System	78
3–13	Organic Backbone of Common Ion Exchange Resins	81

3–14 Selectivity of a Strong Acid Cation Resin 82
3–15 Selectivity of a Strong Base Anion Resin. 86
3–16 Forced Draft Aereator 90
3–17 Cocurrent Demineralization 91
3–18 Cocurrent Demineralization after Regeneration . . . 93
3–19 Countercurrent Demineralization 93
3–20 Typical Caustic Regeneration System 94
3–21 Typical Acid Regeneration System 95
3–22 Ion Exchange Vessel Internals 96
3–23 SAC Exchanger Water Quality during
       a Normal Service Run 97
3–24 SBA Exchanger Water Quality during
       Normal Operation 97
3–25 Common Distributor Arrangement I 99
3–26 Common Distributor Arrangement II 101
3–27 Twin-bed Ion Exchange System Piping Outline . . . 102
3–28 Conventional Filtration 102
3–29 Filtration Spectrum 103
3–30 Spiral-wound Membrane Structure 105
3–31 Generic Outline of a RO Pressure Vessel 106
3–32 Two-Stage RO Schematic 110
3–33 Two-pass RO Schematic 112
3–34 Skid-mounted Reverse Osmosis Unit 113
3–35 EDR with Power Off 121
3–36 EDR with Power On. 121
3–37 The EDI Process 123
3–38 Flow Diagram of a Makeup Water System 126

4–1   Possible Steam Generating Network for
      a Co–generating Facility. 140
4–2   Condenser Tube Map Showing
      Air Removal Compartment 145

LIST OF FIGURES

4–3	"A" Boiler Design	159
4–4	"D" Boiler Design	159
4–5	"O" Boiler Design	159
4–6	Water Circulation in a Drum Boiler	160
4–7	Chimney Boiling in a Deposit	161
4–8	Common Steam Separating Scheme	167
4–9	Relative Solubility of Compounds in Steam	170
4–10	Deposition Pattern in Low Pressure End of a Turbine	171
5–1	A Common Iron Corrosion mechanism	186
5–2	Hydroquinone	189
5–3	Carbohydrazide	189
5–4	Methyl Ethyl Ketoxime	190
5–5	Amine Distribution Ratios vs. Temperature	195
5–6	Normal Injection Points of H_2O_2 into OT Systems	196
5–7	Solubility of Magnetite in Ammonia	197
5–8	Surface Films in Deoxygenated vs. Oxygenated Treatments	198
5–9	Corrosion Characteristics of Carbon Steel vs. pH	205
5–10	Coordinated Phosphate Control Diagram	207
5–11	Control Curves for Coordinated Phosphate Treatments	208
5–12	Structure of EDTA	212
5–13	Change in Silica Volatility vs. Pressure	219
5–14	Allowable Drum Silica Concentrations vs. Pressure	220
5–15	Chemical Feed System	223
6–1	Recommended Sampling Points in a Steam Generation System	226
6–2	Recommended Liquid Sample Nozzle	242
6–3	Multiport Steam Sample Tap	243
6–4	Nozzle for Isokinetic Steam Sampling	244

6–5	Converting Steam to Condensate	246
6–6	Conditioning Rack (front)	248
6–7	Conditioning Rack (back)	249
6–8	Conditioning Configuration	250
6–9	Conditioning System with Controller	251
7–1	Cooling Tower	267
7–2	Fill Material in Cooling Towers – 1	268
7–3	Fill Material in Cooling Towers – 2	268
7–4	Corrosion Reaction of Iron by Oxygen	272
7–5	Corrosion Rates of Steel by Season	286
7–6	Tolyltriazole	290
7–7	Butylbenzotriazole	291
7–8	Nomograph to Calculate LSI	293
7–9	Hydroxyethylidene Diphosphonic Acid	290
7–10	Polyacrylate	291
7–11	Dichlorohydantoin	299
7–12	Chrolinated Isocyanuarate	299
7–13	Dissociation of HOBr to OBr^- and H^+	300
7–14	Bromo-chloro Hydantoin	301
7–15	Isothiozolone Structures	305
7–16	Quaternary Amines	305
7–17	Bromonitropropanediol (BNPD)	305

TABLES

1–1	Weights of Atomic Particles	3
1–2	The 92 Natural Elements	4
1–3	Oxidation Reduction Potentials	29
2–1	Relationships: pH to Hydrogen	40
2–2	Major Components of the Earth's Atmosphere	44
2–3	Most Common Elements in the Earth's Crust	46
2–4	Chemistry of Selected Water Supplies	48
2–5	Chemistry of the Missouri River at Kansas City	49
2–6	$CaCO_3$ Equivalency Table for Common Ions	51
2–7	Water Analysis: Ions as Calcium Carbonate	52
2–8	Solubilities of Mineral Salts	53
2–9	Solubility Product Constants for Common Minerals	55
3–1	Filter Media Sizes	71
3–2	Strong Acid Cation Exchanger Specifications	83
3–3	Ion Exchanger Hydraulic Performance Guidelines	85
3–4	Strong Base Anion Exchanger Specifications	86
3–5	Common Demineralizer Configurations	87
3–6	Pre-treatment Filter Media Specifications	127
4–1	Major Dissolved Ions	146
4–2	Solubility vs. Temperature	156

4–3	Most Common Boiler Deposits	163
4–4	Effects of Boiler Deposits on Heat Transfer	164
4–5	Common Silicate Deposits in Turbines	174
4–6	Heat Tolerance of common Boiler Materials	178
4–7	Composition of Iron- and Nickel-based Used in Power Plant Applications	179
5–1	Most Popular pH-Conditioning Compounds	194
5–2	Feedwater Guidelines	199
5–3	Water Chemistry Guidelines from ASME – 1	200
5–4	Water Chemistry Guidelines from ASME – 2	202
5–5	Water Chemistry Guidelines from ASME – 3	203
5–6	Recommended Steam Purity Guidelines – 1	217
5–7	Recommended Steam Purity Guidelines – 2	217
5–8	General Guidelines for Turbine Contaminants	218
6–1	Steam Turbine Guidelines — 1	238
6–2	Steam Turbine Guidelines — 2	239
6–3	Steam Purity Requirements for Conventional Turbines and Combustion Turbines	240
6–4	Condenser Performance Data	253
6–5	Condenser Tube Correction Factors	262
6–6	Cooling Water Correction Factors	263
7–1	Recommended Dosage Ranges for Corrosion Inhibitor Programs	289
7–2	Comparison of Water Cooling Scale Indices	295
7–3	Relationships of HCO_3, CO_3, and OH to P and M Analyses	298
7–4	Guidelines for Scale Inhibitor Residuals	301
7–5	Typical Dosages for Foulant Treatments	304

AUTHOR'S FOREWORD

I have visited and worked at manufacturing plants in which great emphasis was placed on process chemistry but water and steam chemistry were more or less neglected. After all, it is the product that makes money, right? Yet, a plant shutdown due to a boiler tube failure or steam chemistry upset could cost millions of dollars in lost production. Some failures have even been known to cause fatalities.

While electric utility personnel probably have, as a whole, been more cognizant of steam/water chemistry issues in past years, the industry is changing. Deregulation has spawned utility mergers and breakups of utilities into separate generating and transmission/distribution facilities. The more open power market has also led to the strong growth of independent power producers (IPP) and co-generators, including many large industrial plants. As part of this transition, many utilities are downsizing to operate with as small a staff as possible. Often, technical groups including chemistry departments are the first downsizing targets. Duties once handled by trained personnel are increasingly becoming the task of operators. Similar situations frequently exist at IPPs, where "lean and mean" is the rule not the exception. As at industrial plants, one significant chemistry upset at a utility can lead to production losses and equipment replacement that may cost millions of dollars. Consider the summer of 1998 for example, when at one point during a July heat wave emergency power was selling for $7,000 per megawatt-hour. Loss of a 100-megawatt unit for one hour would incur a financial setback of $700,000!

This book is designed to serve two groups, one being the personnel at utilities, IPPs, and steam-generating industries, who may not have much experience with water/steam chemistry but find themselves in charge of chemistry monitoring or programs. The other group consists of chemical and mechanical engineering undergraduate students who may wish to use this text as a supplement to their courses on steam plant or industrial operation. Today's jobs call for workers that can wear several hats, in which a fundamental knowledge of steam generation chemistry could be very helpful. In my career, I have had to work on steam-generation chemistry projects at electric utilities, a cellophane manufacturing plant, an organic chemical manufacturing firm, a petrochemical plant, and a pharmaceutical manufacturer. At many of these locations, plant personnel had only a partial idea of boiler water and steam chemistry concepts.

Chapter 1 provides an overview of chemistry fundamentals that will help the reader understand the general chemistry of elements and compounds most common to a steam generating system. Chapter 2 discusses some of the unique properties of water, and its natural chemistry. These chapters are not mandatory for understanding the remainder of the book, but they provide building blocks for the remaining material. The succeeding chapters specifically address steam generation chemistry issues, including external treatment of makeup water to remove contaminants, the effects of contaminants on steam generation chemistry, internal chemical treatment programs, sampling and monitoring of water/steam chemistry, and a final chapter about cooling water chemistry. Included in the internal water treatment and sampling chapters are guidelines on water chemistry parameters that have been developed over the years by such organizations as the Electric Power Research Institute, the American Society of Mechanical Engineers, the Heat Exchange Institute, and others.

I have tried to be practical and provide the reader with fundamental concepts, but without extravagant detail. Admittedly, chemistry can be a dry subject if not presented properly, and advanced texts can sometimes serve as a cure for insomnia. It is my hope that this book will provide the reader with information, ideas and solutions that he or she can immediately apply. Many

Author's Forward

examples are taken from my 20 years of experience in the utility, consulting, and manufacturing industries.

Lastly, I would be grateful for any comments on how to improve subsequent editions of the book. If you have any experiences that would make worthwhile additions to a second edition of the book, please forward them to PennWell, who will in turn send them to me. Much of this book is based on actual experience, and I would love to add to my collection, much the same as Scott Adams, the creator of Dilbert, does.

INTRODUCTION

Water. We all know it as one of the two fundamental life-supporting chemicals on the planet. Utility and industrial personnel also know that water in both liquid and gaseous forms is the primary operating or heat-transfer fluid for power generation and other industrial processes. Sometimes less well known is the fact that the quality of water in a steam-generating system can have a dramatic impact on equipment performance and life expectancy. For example, in a high-pressure boiler, the pH must be maintained within or near an alkaline range of 9 to 10 to prevent serious corrosion. Yet, cases are not uncommon where a system upset caused the boiler water pH to drop well below this range and even below the neutral value of 7.0. Such events can cause rapid corrosion, and boiler tubes that should last for years have been known to fail within weeks or even days!

This book is designed to serve as a guide for steam plant operators and engineers, and engineering students, regarding important concepts of steam generation chemistry. This chapter provides a basic introduction to general chemistry, while the second explores the chemistry of water. These fundamentals will help any readers with minimal chemistry training to better understand the remaining chapters.

GENERAL CHEMISTRY 1

Important Physical Concepts: Electric Charge and Gravity

Forces govern our physical world. Some of the most common, readily-observable forces include gravity, magnetism, and friction. Two forces that are of importance to this discussion are gravity and especially electrical charge. The equations for both are similar.

Force of gravity

$$F = G \cdot \frac{m_1 \cdot m_2}{r^2} \qquad [Eq.\ 1\text{--}1]$$

Force of Electrical Charge

$$F = C \cdot \frac{q_1 \cdot q_2}{r^2} \qquad [Eq.\ 1\text{--}2]$$

In the first equation, G is a value known as the gravitational constant, m_1 and m_2 are the masses of two bodies, and r is the distance between them. In the second equation, C is also a constant

1

(the Coulomb constant), q_1 and q_2 are the electrical charges of the attracting bodies, and again r is the distance between the two. I do not intend for this book to serve as a lecture on physics, but these equations illustrate important fundamental principles that are helpful for understanding chemical interactions.

First, the attractive force (or repulsive force in many cases) between bodies is directly proportional to charge and mass. Second, the force is inversely proportional by the power of two to the distance between the bodies. A clear example of the effects of distance and charge is evident when atoms bond to form molecules. Only the outer electrons are available for bonding, in large part because repulsive forces between nuclei become too great to allow atoms to approach very closely. The effect of distance is visually observable in tides. Both the sun and moon influence the earth's tides, but the force from the moon is more powerful, even though the sun is enormously more massive. The sun's much greater distance (r), when squared, accounts for this difference.

Atoms and Molecules: Nature's Building Blocks

Although matter consists of many subatomic particles, the basic components for understanding everyday chemistry are atoms and molecules and their building blocks, protons, neutrons, and electrons.

Atomic Structure and Size

As many of you may recall from high school or college science courses, an atom consists of a nucleus containing positively charged protons and uncharged neutrons, with negatively charged electrons in constant orbit. The weights of the individual particles are very small (Table 1–1).

An amazing feature about atoms is how tiny they are yet how much empty space is contained between the nucleus and the electrons. The proton is only about 10^{-13} centimeters in diameter, with the neutron being fractionally larger and the electron over 1800 times smaller. Yet, the atom

CHAPTER 1: GENERAL CHEMISTRY

Weights of Atomic Particles

Particle	Weight
Proton	1.6725×10^{-24} grams*
Neutron	1.6748×10^{-24} grams
Electron	9.109×10^{-28} grams

* *For those readers unfamiliar with scientific notation, please refer to Learning Aide 1–1 at the end of this chapter.*

Table 1–1 *Atomic Particle Weights*

as a whole has a diameter 10,000 or more times that of the nucleus because the electrons orbit at great distances relative to the nuclear size. Astronomical observations provide an interesting example of the empty space in and between atoms on earth as compared to other bodies. Consider a dense substance on our planet, such as liquid mercury. A five-gallon bucket of this material weighs 565 pounds. As Marty in the *Back to the Future* movies might say, "That's heavy, doc!" Yet, consider a neutron star, in which gravitational forces compress atoms until protons and electrons combine to form neutrons. One teaspoon of this material weighs 500 million tons!

Atomic Number, Atomic Mass, and Isotopes

On earth, the chemistry that governs most chemical processes involves bonding of atoms through electron interaction. Very common examples are all around. If you go to a wiener roast and watch the logs burn, you see light and feel heat produced by the reaction, through electron bonding, of carbon atoms in the wood with oxygen. The rusting process is the conversion of elemental iron to iron oxide through electron bonding of iron and oxygen. Plastics are made up of giant molecules known as macromolecules that take advantage of the unique electron-bonding characteristics of carbon. The list is enormous. Everything we use, touch, and see consists of atoms linked by electrons.

The 92 Natural Elements

Element	Symbol	Atomic No.	Atomic Mass	Element	Symbol	Atomic No.	Atomic Mass
Actinium	Ac	89	227.028	Neodynium	Nd	60	144.24
Aluminum	Al	13	26.982	Neon	Ne	10	20.179
Antimony	Sb	51	121.75	Nickel	Ni	28	58.69
Argon	Ar	18	39.948	Niobium	Nb	41	92.906
Arsenic	As	33	74.92	Nitrogen	N	7	14.007
Astatine	At	85	(210)	Osmium	Os	76	190.2
Barium	Ba	56	137.33	Oxygen	O	8	15.999
Beryllium	Be	4	9.012	Palladium	Pd	46	106.42
Bismuth	Bi	83	208.98	Phosphorus	P	15	30.974
Boron	B	5	10.811	Platinum	Pt	78	195.08
Bromine	Br	35	79.904	Polonium	Po	84	(209)
Cadmium	Cd	48	112.41	Potassium	K	19	39.098
Calcium	Ca	20	40.078	Praseodymium	Pr	59	140.908
Carbon	C	6	12.01	Protactinium	Pa	91	231.036
Cerium	Ce	58	140.12	Promethium	Pm	61	(145)
Cesium	Cs	55	132.905	Radium	Ra	88	226.025
Chlorine	Cl	17	35.453	Radon	Rn	86	(222)
Chromium	Cr	24	51.996	Rhenium	Re	75	186.21
Cobalt	Co	27	58.93	Rhodium	Rh	45	102.906
Copper	Cu	29	63.546	Rubidium	Rb	37	85.468
Dysprosium	Dy	66	162.50	Ruthenium	Ru	44	101.07
Erbium	Er	68	167.26	Samarium	Sm	62	150.36
Europium	Eu	63	151.96	Scandium	Sc	21	44.96
Fluorine	F	9	18.998	Selenium	Se	34	78.96
Francium	Fr	87	(223)	Silicon	Si	14	28.086
Gadolinium	Gd	64	157.25	Silver	Ag	47	107.868
Gallium	Ga	31	69.723	Sodium	Na	11	22.990
Germanium	Ge	32	72.59	Strontium	Sr	38	87.62
Gold	Au	79	196.966	Sulfur	S	16	32.056
Hafnium	Hf	72	178.49	Tantalum	Ta	73	180.948
Helium	He	2	4.003	Technetium	Tc	43	(98)
Holmium	Ho	67	164.93	Tellurium	Te	52	127.60
Hydrgoen	H	1	1.008	Terbium	Tb	65	158.925
Indium	In	49	114.82	Thallium	Tl	81	204.383
Iodine	I	53	126.9044	Thorium	Th	90	232.038
Iridium	Ir	77	192.22	Thullium	Tm	69	168.93
Iron	Fe	26	55.847	Tin	Sn	50	118.71
Krypton	Kr	36	83.80	Titanium	Ti	22	47.88
Lanthanum	La	57	138.906	Tungsten	W	74	183.85
Lead	Pb	82	207.2	Uranium	U	92	238.029
Lithium	Li	3	6.941	Vanadium	V	23	50.941
Lutetium	Lu	71	174.967	Xenon	Xe	54	131.29
Magnesium	Mg	12	24.305	Ytterbium	Yb	70	173.04
Manganese	Mn	25	54.938	Yttrium	Y	39	88.906
Mercury	Hg	80	200.59	Zinc	Zn	30	65.39
Molybdenum	Mo	42	95.94	Zirconium	Zr	40	91.224

Table 1–2 *The 92 Natural Elements*

CHAPTER 1: GENERAL CHEMISTRY

The earth is made up of 92, natural chemical elements (Table 1–2), and scientists have artificially produced another 15 or so. The element or atomic number is determined by the protons in the nucleus. For example, hydrogen with one proton has an atomic number of 1. Oxygen has eight protons so its atomic number is 8. Iron is element 26; silver is element 47, and so forth.

In every element above hydrogen, and even in some forms of hydrogen, the nucleus contains neutrons. Helium, the second element, has a nucleus consisting of two protons and two neutrons. The most common form of the heaviest natural element, uranium-238, has 92 protons and 146 neutrons. Because the proton and neutron are so close in mass, and are much larger than the electron, an atom's mass number can, in general, be determined by adding the protons and neutrons. Thus, for helium the element number is two, but the mass number is four. Precise atomic mass values are based on carbon-12, which has six protons and six neutrons, and by definition is considered to have an atomic mass of 12.0000... All of the atomic weights listed in Table 1–2 come from this standard.

Atoms with the same atomic number but different atomic mass are known as isotopes. Probably the most famous isotope is uranium-235 (92 protons, 143 neutrons), which was used in the first atomic weapon. Another classic example is carbon-14. The vast majority of carbon atoms are carbon-12, which, as we just seen, contain six protons and six neutrons. However, cosmic radiation bombarding earth converts some nitrogen-14 atoms (seven protons and seven neutrons) into carbon-14 with six protons and eight neutrons. Carbon-14 continually moves through the biological life cycle just like its normal carbon-12 relative, but being an unstable compound eventually decays radioactively to nitrogen. The half-life is 5760 years. When organisms die, carbon no longer enters the body and the carbon-14 content decreases through radioactive decay. Scientists can determine the age of many ancient materials by comparing the carbon-14 content versus the steady concentration found in living matter.

The isotopic effect on the average atomic mass of each element is incorporated in Table 1–2. That is why carbon as a whole has an atomic weight of 12.011 rather than 12.0000. Enough isotopes are present in nature to increase the average mass by the fraction shown.

More About Electrons

An atom of any element in a normal state contains an equal number of electrons and protons. Thus, as atomic number increases so does the corresponding number of electrons. The manner in which electrons align around atomic nuclei is predicted by quantum mechanics, which is a special branch of chemistry and physics. The math is very complex and is beyond the scope of this book. However, the predictions of quantum mechanics fit observed data very well, and allow for practical understanding of the behavior of elements. Quantum mechanics predicts, and experimentation has shown, that electrons reside in discrete energy levels around the atom, and that only a select number of electrons can fill any one level. Quantam mechanics does not predict the exact locations of electrons at any particular moment, but rather determines regions of electron density. These regions are known as orbitals. Orbital theory is quite detailed, so we will not examine it in depth. What we will look at are orbitals as they relate to the shell theory of electron configurations. As we shall see, the number of orbitals and electron shells increase with atomic number. Shell energies and the number of electrons within each shell greatly influence the behavior of the atom with regard to its own properties and bonding with other atoms.

Electrons at one energy level can be induced to jump to another energy level, if given the correct type and sufficient amount of energy. These energy levels are quite precise and are said to be quantized. When the electron returns to its normal state, it gives off energy, which sometimes may appear as visible light. A practical example of this will be illustrated later.

Chemistry Basics and The Periodic Table

Although some similarities exist among various groups within the 92 natural elements, each element exhibits unique properties due to size, electron configuration, nuclear charge, and other properties. This makes

CHAPTER 1: GENERAL CHEMISTRY

Periodic Table
with Atomic Number, Atomic Mass, and Orbital Configurations

Period	Orbitals	IA	IIA	IIIB	IVB	VB	VIB	VIIB	VII			IB	IIB	IIIA	IVA	VA	VIA	VIIA	O
1	1s	1 H 1.008																	2 He 4.003
2	2s, 2p	3 Li 6.941	4 Be 9.012											5 B 10.811	6 C 12.01	7 N 14.007	8 O 15.999	9 F 18.998	10 Ne 20.179
3	3s, 3p	11 Na 22.990	12 Mg 24.305											13 Al 26.982	14 Si 28.086	15 P 30.974	16 S 32.056	17 Cl 35.453	18 Ar 39.948
4	4s, 3d, 3p	19 K 39.098	20 Ca 40.078	21 Sc 44.96	22 Ti 47.88	23 V 50.941	24 Cr 51.996	25 Mn 54.938	26 Fe 55.847	27 Co 58.93	28 Ni 58.69	29 Cu 63.546	30 Zn 65.39	31 Ga 69.723	32 Ge 72.59	33 As 74.92	34 Se 78.96	35 Br 79.904	36 Kr 83.80
5	5s, 4d, 5p	37 Rb 85.468	38 Sr 87.62	39 Y 88.906	40 Zr 91.224	41 Nb 92.906	42 Mo 95.94	43 Tc (98)	44 Ru 101.07	45 Rh 102.906	46 Pd 106.42	47 Ag 107.868	48 Cd 112.41	49 In 114.82	50 Sn 118.71	51 Sb 121.75	52 Te 127.60	53 I 126.904	54 Xe 131.29
6	6s, 4f, 5d 6p	55 Cs 132.905	56 Ba 137.33	57 La 138.906	72 Hf 178.49	73 Ta 180.948	74 W 183.85	75 Re 186.21	76 Os 190.2	77 Ir 192.22	78 Pt 195.08	79 Au 196.966	80 Hg 200.59	81 Tl 204.383	82 Pb 207.2	83 Bi 208.98	84 Po (209)	85 At (210)	86 Rn (222)
7	7s, 5f, 6d	87 Fr (223)	88 Ra 226.025	89 Ac 227.028															

Lanthanide Series	58 Ce 140.12	59 Pr 140.908	60 Nd 144.24	61 Pm (145)	62 Sm 150.36	63 Eu 151.96	64 Gd 157.25	65 Tb 158.925	66 Dy 162.50	67 Ho 164.93	68 Er 167.26	69 Tm 168.93	70 Yb 173.04	71 Lu 174.967
Actinide Series	90 Th 232.038	91 Pa 231.036	92 U 238.029	93 Np 237.05	94 Pu (244)	95 Am (243)	96 Cm (247)	97 Bk (247)	98 Cf (251)	99 Es (252)	100 Fm (257)	101 Md (258)	102 No (259)	103 Lr (260)

Fig. 1–1 *The Periodic Table*

the study of chemistry fascinating but at the same time very complex. Thousands of books have been written about chemistry in general or specific aspects thereof. In this book, I wish to highlight aspects of the elements that have a bearing on their behavior in water and, most particularly, in steam generating systems. To do so, I will use the tool known as the periodic table.

Figure 1–1 shows the classic arrangement of the first 103 elements discovered. Many of you may recognize it as the periodic table. The scientist, Mendeleev, and others who followed him developed the periodic table as a means of organizing elements into logical groups. Indeed, one can find some very good logic in the organization of the table. However, atomic properties are so variable that the behavior of many elements must be explained on an individual basis.

The periodic table is divided into both rows and columns. Vertical columns are known as families, and their members often show similar chemical properties. The horizontal rows are known as periods, and are arranged according to electron shell level. Several general trends of the periodic table reveal much about the behavior of the elements. Important points to remember about atoms and the periodic table include the following:

- Each new row in the periodic table represents a new electron shell. Each new electron shell resides at a higher energy level than the previous shell.

- In many non-metallic bonds, each atom shares an electron with its neighbor and vice versa. Thus, each bond consists of a shared pair of electrons between atoms. Some atoms, carbon in particular, can form double and triple bonds.

- Each electron shell is made up of an increasing number of atomic orbitals. The first shell, in which hydrogen and helium are the only two elements, has one orbital defined by quantam mechanics as a "1s" orbital. The second row has an electron shell made up of four orbitals, designated as one "2s" and three "2p" orbitals. Each orbital can only hold a maximum of two electrons. Thus, the maximum number of bonds possible in this row is four. The third row has a "3s" and three "3p" orbitals, but the chemistry is much different than the second row. The fourth and fifth rows have not only "s" and "p" orbitals, but also five "d" orbitals. Elements 21–30 and 39-48 are known as the transition metals, and it is in these sequences that the "d" orbitals fill with electrons. Some heavier third row elements also exhibit "d" orbital properties, which is one reason why third row chemistry is different from second row chemistry. At element 57 and again at element 89, another set of orbitals, the "f" orbitals become available.

- As one proceeds vertically downward through the table, atomic size increases but the ionization enthalpy (energy needed to remove outer electrons) decreases.

- As one proceeds horizontally along a row, the atomic size decreases and the ionization enthalpy generally increases.

- Most of the elements are metals. In the majority of families whose early members (boron, carbon, nitrogen, and oxygen) are not metals, metallic behavior increases as one proceeds vertically through the table.

Many of these trends will become apparent as we examine the table. Let's start with the first row, which has only two elements. This is because the first electron shell has only one orbital, the "1s" orbital, and can only hold two electrons. The first element is hydrogen, with one electron. Following is helium with the maximum of two electrons. Atoms with filled electron shells are in general very stable, and observation proves this. Helium is an inert gas, and is the most non-reactive element of all. The filled shell, with a small central charge holding the electrons tightly (remember Equation 1–2), is extraordinarily stable. The stability of a filled shell also explains why hydrogen in its elemental state always exists as the paired molecule, H_2. Two hydrogen atoms, with one electron apiece, will bond together to reach a filled shell structure. It takes a significant amount of energy to break this bond (activation energy), because of the stability of the filled shell. However, if enough energy is supplied to break the hydrogen molecule, the atoms will react violently with other elements to re-establish a bond. The Hindenburg dirigible disaster in 1937 is the classic example of hydrogen's release of energy when bonding. An electrical spark from a mooring tower provided enough activation energy to break apart some hydrogen molecules. Their subsequent reaction with oxygen released more than enough energy to sustain and increase the process.

The second row in the periodic table consists of eight elements from lithium to neon. The second shell, which contains a maximum of eight electrons at neon, is made up of the "2s" and three "2p" orbitals. Element 3, lithium, shows a very interesting trend, a trend which plays an important part in water chemistry. The first two electrons of a lithium atom reside in the filled, "1s" shell, so the third electron is added to the new shell. This electron is weakly attracted to the nucleus because the second shell is at a higher energy level than the first, and because the electrons in the filled inner shell partially screen the third electron from the nuclear charge. Thus, this outer electron is highly influenced by external forces and is readily available for bonding with other atoms. This trend continues and actually increases down this family (known as the alkalis) of elements, sodium (Na), potassium (K), rubidium (Rb), and cesium (Cs). The increased ability of members within the family to donate the outside electron is primarily due to two factors. One, the outer electrons are further from the nucleus, which increases r in Equation 1–2. Second, each succeeding element contains more lower-energy electrons, which shield or screen the outside electron from the nucleus. Because the outer electron in the alkalis is held so loosely, these elements have a strong tendency to ionize, wherein the electron is essentially captured by another element, leaving the alkalis as positively-charged ions (cations) with a $^{+}1$ charge. The tendency of the alkalis to ionize has a profound effect on their behavior in water, as will be shown. Sodium and to a lesser extent potassium often make up a considerable portion of dissolved solids in water. The alkalis in elemental form are metals, which by definition includes the ability to conduct electricity. Metallic properties occur in elements that contain partially filled orbitals, in which one or more electrons are free to move about.

The alkalis are extremely reactive, and in elemental form must be stored in a non-reactive fluid or atmosphere, as otherwise they will combine violently with air or moisture. A professor of mine told a story about his early college days, in which he met a fellow undergraduate student who had been kicked out of a previous university for throwing a half-pound of elemental sodium into the school's swimming pool. The reaction destroyed the tiles and concrete wall at the end of the pool into which he had thrown the sodium!

The next family of elements is known as the alkaline earths, and they exhibit somewhat similar properties, including metallic behavior, to the alkalis. In this family, the atoms contain two electrons that are partially screened by the filled shell below. These atoms, although not quite as reactive as the alkalis, also readily donate electrons, and, with the exception of beryllium, form cations with a +2 charge. The alkaline earths are beryllium, magnesium, calcium, strontium, and barium. These elements also play a very important part in water chemistry. Magnesium and calcium are the major hardness ions, and they will precipitate with a variety of anions (negatively-charged ions) to form scale deposits in boilers, cooling water systems, makeup treatment equipment, and even home plumbing. Strontium and barium are not usually found in high concentrations in natural water supplies, but where the water may become concentrated, such as the final stages of a reverse osmosis unit, the ions will combine with the sulfate anion (SO_4^{-2}) to form troublesome deposits.

As we move along the second row, other trends become apparent. Most importantly, the tendency to ionize into a positive state greatly decreases and then disappears. Boron, with three outer electrons, does not readily form cations, and it does not exhibit metallic properties. Boron shows a very strong affinity for oxygen, and is always found in nature combined with this element. Once, during a college chemistry lab, I accidentally released a cloud of boron trifluoride (BF_3). It immediately formed a fuming vapor directly in front of me. I watched as this fuming ball rose to the ceiling and then disappeared into a ceiling exhaust vent. Fortunately, my instructor did not observe this unplanned experiment, and I proceeded with the assigned experiment as planned.

Carbon is an extraordinarily unique atom, and life as we know it would not exist without its special bonding properties. At least three aspects of carbon make it so important. One, the four electrons that comprise the second shell of a carbon atom are all available for bonding. Secondly, carbon will bond with other carbon atoms in chains or rings to form very stable structures. Diamonds are carbon atoms connected to each other by single bonds in a series of tetrahedrons. Third, carbon can form multiple bonds with other atoms, and especially with other carbon atoms. For example, acetylene, the gas used for metal cutting and welding, is a two-carbon molecule in which the carbon atoms are triply bonded. Graphite, used in pencil lead and as a lubricant, consists of layers of carbon rings that contain alternating single and double bonds. About a decade or so ago, researchers discovered carbon molecules that resemble molecular soccer balls, with 60 atoms all linked together in a spherical shape. These "Bucky Balls" continue to amaze scientists.

And, of course, there are the fossil fuels. Coal, the byproduct of ancient vegetation, is a complicated structure containing many carbon rings and chains along with smaller amounts of hydrogen, oxygen, nitrogen, sulfur, and other inorganic compounds. Oil is the carbon-based remnants of ancient sea organisms.

Life depends upon carbon-carbon bonding. Hundreds of thousands of carbon compounds have been identified, and a whole branch of chemistry, known as organic chemistry, is devoted to these compounds. Natural organics, produced from decaying vegetation, can have an impact on water treatment and steam-generation chemistry, as will be outlined later.

The most important carbon-containing compounds with regard to industrial water chemistry and water treatment are carbon dioxide (CO_2), bicarbonate (HCO_3^-), and carbonate (CO_3^{-2}). Carbon dioxide is, of course, a natural component of the atmosphere. CO_2 is absorbed by moisture, which increases the acidity of precipitation and influences the behavior of natural water. Carbon dioxide's relatives, bicarbonates and carbonates, are common constituents of water supplies and minerals, respectively. Carbonate deposits, primarily calcium and magnesium carbonate, are found throughout the world as limestone. Limestone is a very important construction material, and among other things is a major ingredient in concrete.

Up to this point in the second row, each electron added to the second shell has been bonding in nature. From here on to the last element in the row, neon, the electrons that are added do not exhibit bonding characteristics.

Proceeding further we come to nitrogen, oxygen, and fluorine, all gases in the elemental state. Here we see the development of anionic tendencies. The maximum number of allowable electrons in the second shell is eight. Each succeeding element begins to approach a filled, and therefore stable, shell configuration. This causes the latter elements in the row to seek electrons to complete the shell configuration. Nitrogen, another basic chemical of life, has five electrons in the outer shell, three of which are available for bonding. Several important nitrogen compounds are based upon the three-bond configuration. Ammonia (NH_3) is one of the most well known nitrogen compounds, and ammonia or organic derivatives known as amines, are almost universally used as pH control chemicals in steam generating systems. Nitrogen in the lower atmosphere exists as N_2 because two nitrogen atoms form a very strong triple bond. The bond is so strong that nitrogen is often used as an inerting material to control or prevent other reactions. For example, nitrogen blanketing of boilers is a common method to prevent oxygen corrosion when the boilers are off-line.

Oxygen exhibits some unique properties. Of oxygen's six outer electrons, only two are available for bonding. Oxygen exists in the atmosphere as the double-bonded molecule O_2, which completes the eight-electron shell for each atom. The double bond makes oxygen relatively non-reactive under normal circumstances. However, oxygen is only two electrons short of a filled shell structure, and if enough activation energy is provided to break the oxygen-oxygen bond, the atoms will readily combine with other elements to form more stable compounds. Combustion is, of course, an example of this process. Dissolved oxygen in water is quite reactive towards iron and many iron-based alloys, and oxygen has historically been one of the most troublesome agents in steam generating systems.

Much of the energy produced throughout the world is consumed in industrial processes that reverse oxygen bonding with other elements. Iron and aluminum production are practical examples. Natural deposits of these metals do not exist in an elemental state, but are almost always combined with oxygen or oxygen-containing anions. A large amount of energy is needed to break these metal-oxygen bonds and convert the metals to elements. That is why recycling of aluminum cans is important.

It takes much less energy to collect and melt the cans than is required to convert aluminum oxide ores to elemental aluminum.

Oxygen is the most common element in the earth's crust and exists in combination with many other elements. Soils and clays are complex materials in which oxygen is bound with silicon in compounds known as silicates. Other common forms of oxygen-bearing materials include sulfates (SO_4) and carbonates (CO_3). These are often found in natural deposits of gypsum ($CaSO_4 \cdot 2H_2O$) and limestone ($CaCO_3$ and $MgCO_3$).

The next to last element in the second row is fluorine. Fluorine, which has a small nuclear size and needs only one electron to complete the electron shell, is the most reactive element of all. Fluorine can be isolated as a pure substance, in which case it exists as F_2, but in nature fluorine always combines with other elements. It is part of the family known as the halogens.

The last element in the second row is neon. Like helium, neon has a completed electron shell and is very stable. Helium, neon, and the other members of the family are known as the noble gases. Neon is not quite as inert as helium because the outer electrons are further from the nucleus and also, screening effects by inner electrons prevent the outer electrons from seeing a full nuclear charge. Neon lighting represents a practical illustration of electron energy states. When an electric charge is passed through neon gas, some of the electrons receive enough energy to jump into a higher state. As the electrons return to their normal energy level they give off light.

An examination of the third row illustrates some of the similarities and differences between members of the same families. The third row has the same number of elements as the second row, and, at first glance, the same number of orbitals, although at a higher energy state. In certain families the elements behave similarly to their earlier counterparts. This is true with the alkalis and the noble gases. Sodium and magnesium have strong cationic tendencies while argon is inert. Chlorine behaves somewhat differently than fluorine due to its larger size, but chlorine is also strongly anionic and seeks an electron. A classic compound that illustrates both the chemistries of the alkalis and halogens is table salt, sodium chloride. As we shall see in the section on ionic bonding, sodium chloride crystals form a rigid lattice with sodium atoms existing essentially in a +1 oxidation state and chlorine atoms in a -1 oxidation state.

Other comparisons between the third row and second show the differences caused by increased atomic size and more complicated electron configurations. Aluminum is very definitely a metal, whereas boron is not. Aluminum is also very electropositive like the alkalis and alkaline earths, and will readily donate electrons when bonding. This is why aluminum always exists in nature as a mineral rather than an element. Aluminum has proven to be a very useful material for construction due to its light weight and the fact that aluminum forms a very stable coating of aluminum oxide which protects the underlying material. Aluminum is a particularly important material of construction in the aircraft industry.

Silicon, which is directly below carbon in the periodic table, can form multiple bonds, but not to the same extent as carbon. Silicon develops very strong bonds with oxygen, and, since oxygen and silicon are the two most common elements in the earth's crust, explains why silicates are the predominant component of soils.

Phosphorous is not found in nature in elemental form but is combined with oxygen in mineral deposits. Sulfur does exist in elemental form but is also often found in mineral deposits as the sulfate and sometimes the sulfide ion (S^{-2}). Fool's Gold is simply iron sulfide. Phosphorous and sulfur are among the first elements in the periodic table that exhibit the ability to form more than four bonds. Sulfur hexaflouride (SF_6) is an insulating material used in the power industry. This multiple bonding occurs through "d" orbitals, which becomes fully pronounced in the next row of the periodic table.

In the fourth row, five additional orbitals (known as the "d" orbitals) become available. These orbitals begin to fill at element 21, scandium, through element 30, zinc. This group of ten elements is known as the transition metals, and includes iron, nickel, and copper. The transition elements are metals because they have partially filled orbitals with one or more loosely-held electrons. Many transition elements are moderately-to-fairly strongly electropositive and will give up electrons. This explains why many of them do not exist in nature as the pure element. Iron is always combined with an anion, usually oxygen, but sometimes carbonate or sulfide. Chromium is typically found in nature as the chromite ore ($FeCr_2O_4$). Chromium is the alloying material in stainless steel. It functions by developing a chromic oxide layer on the surface of the steel that protects the underlying iron from rust formation. Dissolved manganese, along with iron, is a common impurity in well water. When exposed to the atmosphere, both convert to oxides that stain plumbing fixtures and laundry. Zinc is the coating on galvanized steel.

Copper is a very important transition metal because of its good electrical conductivity, excellent heat transfer capabilities, and relative stability to corrosion. Copper alloys have been a prime material of construction for heat exchangers. Other transition metals that are often used as alloys for steam plant materials include nickel and a fifth-row-transition metal, molybdenum. Somewhat recently, titanium has become rather popular as a material of construction for heat exchanger tubes.

Another important element in the fourth row is bromine. Elemental bromine (Br_2) exists as a liquid. Like chlorine, it is reactive and seeks electrons. Bromine is becoming increasingly popular as a microbiocide, and is replacing chlorine in many applications. More about this in chapter 7.

The fifth row shows a very similar trend of "d" orbital filling, and has another set of transition elements from elements 39 through 48. This group includes silver.

In the sixth and seventh rows, yet another group of orbitals becomes available. These are the "f" orbitals. These orbitals fill from elements 58 through 71 and 90 through 103. Very little of the chemistry outlined in this book involves these elements, so we will not discuss them further.

Other aspects of the periodic table will come to light as we examine more features about atoms and molecules in subsequent sections.

Valence and Electronegativity

We have already seen that the alkalis tend to give up one electron to form a single-charged cation, and the alkaline earths give up two electrons to form a double-charged cation. The halogens, fluorine, chlorine, bromine, and iodine exhibit the opposite effect; they strongly seek an electron to complete an electronic shell. A common term for the electronic state in which an element exists within a compound is valence. Consider again sodium chloride, NaCl. The valence of sodium in this compound is +1 while that of chlorine is -1. In elemental chlorine, which exists as Cl_2, the valence of each chlorine atom is zero, since the two atoms are identical. As this indicates, the valence of an element depends upon how it is combined with other elements. Many elements, and particularly the transition metals, have several valences. Iron is a good example. In human blood, iron exists in a +2 valence state. This gives blood its bright red color. When blood is exposed to air it turns brown. The color change is caused by the reaction of iron with oxygen and its conversion from a +2 to +3 valence state. Rust formation illustrates an even greater change in the valance state of iron. This is a two-step process whereby iron in the elemental state (valence of zero) is first converted to a +2 state and then a +3 state, as oxygen grabs available electrons.

> ## Important Terminology
> When an element gives up electrons during a reaction it is said to be *oxidized;* when it accepts electrons it is said to be *reduced.* The term for a reaction where electrons are transferred is *redox reaction.* Redox reactions are fundamental for understanding corrosion, and we will examine them in greater detail later in this chapter.

Another useful concept, which provides an indication of the electron attraction capabilities of an atom, is electronegativity. Electronegativity is a measure of an atom's ability not only to seek other electrons, but also to hold on to its own. Several methods have been developed to calculate electronegativities, and Figure 1–2 shows the values predicted by one of these methods. Higher values indicate greater electronegativity while lower values indicate the opposite.

As the table illustrates, fluorine is the most electronegative element of all, with a calculated value of 4.1. The alkalis have the lowest values. We would expect this based on the position of the elements in the periodic table. While electronegativities cannot be used as an absolute predictive model for bond formation, they are helpful in determining the nature and polarity of chemical bonds. In the hydrogen fluoride (HF) molecule,

Fundamentals of Steam Generation Chemistry

Electronegativities

IA	IIA	IIIB	IVB	VB	VIB	VIIB	VIII			IB	IIB	IIIA	IVA	VA	VIA	VIIA	O
H 2.1																	He
Li 0.97	Be 1.47											B 2.01	C 2.50	N 3.07	O 3.50	F 4.10	Ne
Na 1.01	Mg 1.23											Al 1.47	Si 1.74	P 2.06	S 2.44	Cl 2.83	Ar
K 0.91	Ca 1.04	Sc 1.20	Ti 1.32	V 1.45	Cr 1.56	Mn 1.60	Fe 1.64	Co 1.70	Ni 1.75	Cu 1.75	Zn 1.66	Ga 1.82	Ge 2.02	As 2.20	Se 2.48	Br 2.74	Kr
Rb 0.89	Sr 0.99	Y 1.11	Zr 1.22	Nb 1.23	Mo 1.30	Tc 1.36	Ru 1.42	Rh 1.45	Pd 1.35	Ag 1.42	Cd 1.46	In 1.49	Sn 1.72	Sb 1.82	Te 2.01	I 2.21	Xe
Cs 0.86	Ba 0.97	La 1.08	Hf 1.23	Ta 1.33	W 1.40	Re 1.46	Os 1.52	Ir 1.55	Pt 1.44	Au 1.42	Hg 1.44	Tl 1.44	Pb 1.55	Bi 1.67	Po 1.76	At 1.90	Rn
Fr 0.86	Ra 0.97	Ac 1.00															

Lanthanide Series	Ce 1.08	Pr 1.07	Nd 1.07	Pm 1.07	Sm 1.07	Eu 1.01	Gd 1.11	Tb 1.10	Dy 1.10	Ho 1.10	Er 1.11	Tm 1.11	Yb 1.06	Lu 1.14
Actinide Series	Th 1.11	Pa 1.14	U 1.22	Np	Pu	Am	Cm	Bk	Cf	Es	Fm	Md	No	

Fig. 1–2 *Electronegativities*

where hydrogen has an electronegativity value of 2.1, fluorine exerts a strong influence on the hydrogen electron. This imparts polarity to the molecule, in which the fluorine develops a partial negative charge and hydrogen a partial positive charge. Polarity is an extremely important concept, and, as we shall see in chapter 2, is a phenomenon that has a great impact upon the behavior of water.

Chemical Bonding / Types of Bonds

It is useful now to briefly examine the various types of chemical bonds. All involve sharing of electrons, but it is the manner in which the electrons are shared that give each bond its own characteristics. The four types of bonds we will look at are ionic, covalent, metallic, and those described by coordination chemistry.

Ionic Bonds

In our examination of atomic properties, we found that some elements are very hungry for an extra electron or two to complete a filled shell configuration. This is especially true with fluorine, oxygen, and chlorine. Conversely, the electropositive elements such as the alkalis and alkaline earths have one or two electrons that are bound quite loosely. When these atoms combine to form molecules, the difference in electronegativity generates bonds that can almost be considered electrostatic in nature rather than a mutual sharing of electrons. The highly electronegative atoms for the most part capture the electron or electrons donated by the electropositive atoms. Thus, an ionic bond can be pictured fairly accurately as the electrostatic union of a cation (positively-charged ion) with an anion (negatively-charged ion). Ionic compounds form crystalline solids with alternating cations and anions. Other elements besides the alkalis and alkaline earths serve as cations in ionic compounds. These include metals, transition and otherwise, that exhibit strong electropositive behavior. Aluminum, iron, chromium, and titanium are examples.

While cations are simple atoms, anions vary in complexity. The simple anions are fluorides, chlorides, oxides, and sulfides. However, many molecular anions exist, and many are found in nature. The most common include carbonate (CO_3^{-2}), sulfate (SO_4^{-2}), nitrate (NO_3^-), and phosphate (PO_4^{-3}). Even more complex anions are based on the silicate structure (Si_xO_y).

The stability of an ionic solid is dependent upon a number of factors, including the strength of the bonds between atoms. One factor that influences bond strength is the charge of the ions. Calcium carbonate forms a stronger lattice than say potassium chloride (KCl) because both the cation (Ca^{+2}) and anion (CO_3^{-2}) are divalent as opposed to the monovalent cations in the simpler salt. Lattice strength is also enhanced when the size of the cations and anions is similar, allowing for tight structuring of the atoms. An example of both size and charge factors can be seen when comparing the bond strength of sodium chloride with calcium fluoride (CaF_2). The lattice strength of CaF_2 is almost three times greater than that of NaCl. Lattice bonding energy influences the dissolution properties of minerals exposed to natural waters and also the strength and tenacity of deposits that form in steam generating systems. We will return to this topic in Chapter 2, when we discuss the ability of water to dissolve minerals.

Covalent Bonding

In many molecules, the electronegativities of the individual atoms are close enough that bonding consists of a mutual sharing of electrons. A good example is the methane molecule (CH_4). Each of the four hydrogen atoms bonds to carbon through a shared pair of electrons, one donated by hydrogen and one by carbon. This bonding allows hydrogen to reach a filled shell level of two electrons and carbon to reach a filled shell level of eight. This type of bond is known as a covalent bond. Millions of covalently bonded compounds are known, and this adds to the complexity of any study of chemistry. One property often exhibited by covalently bonded molecules is their lack of attraction for and by surrounding molecules. In a true covalent molecule, such as the diatomics hydrogen, nitrogen, oxygen, fluorine, and others, the 100% mutual sharing of electrons makes these molecules non-polar and they show little attraction towards one another. This in some measure explains why all are gases. Even chlorine, which is heavier than air, exists as the gas Cl_2.

Metal Bonding

As we have seen, many elements in the periodic table are metals. The atoms in a metal form a closely packed array with their neighbors. Three basic structures are known—cubic, hexagonal close packed, and body centered cubic. We will not examine these structures in any more detail other than to point out that depending upon the structure, each atom may have eight or twelve neighbors. Metallic bonding introduces a complexity to bonding theory, because insufficient electrons are available to form a covalent bond between an atom and all its neighbors. Scientists developed the band theory to explain metallic bonding. In very simplified terms, band theory says that electrons in metals are not confined to particular regions, but may spread out (become delocalized) in an energy band that extends throughout the metal. This theory correlates with the observed behavior of metals, which includes high-electrical conductance, high thermal conductance, and for many metals and especially the transition elements, high strength and ductility.

Coordination Chemistry

The latter elements of the third row and the transition metals of succeeding rows have "d" orbitals (and in the sixth and seventh rows, "f" orbitals) available for bonding. This allows the elements to bond with more than four other atoms, which is the maximum bond order in the second row. We have already seen that sulfur will bond with fluorine to form SF_6. The transition metals in particular display this type of bonding, and many four, five, six, and even eight coordinate structures are known. This chemistry is not exceptionally important for the remainder of this book, but at least one example of it will appear later.

Chemical Reactions

The final major section of this chapter discusses chemical reactions. A general understanding of reaction mechanisms serves as a building block for understanding chemistry in steam generating systems.

Chemical reactions are energy driven. When a reaction occurs, the reactants are seeking a lower energy state. Most reactions give off heat (exothermic reactions) although a few will absorb heat (endothermic) during the process. As we learned with the example of the hydrogen-oxygen reaction discussed earlier, sometimes an activation energy must be supplied to break existing bonds so that the desired (or undesired as in the case of the Hindenburg) reaction may proceed. Although many millions of chemical compounds exist, the mechanisms by which reactions occur can be broken down into four basic categories. These are acid-base reactions, addition and elimination reactions, substitution reactions, and oxidation-reduction reactions. We will also look at reactions that do not proceed to completion but reach an equilibrium. First, I want to introduce a concept that will allow us to examine chemistry in quantities that can be comprehended.

The Mole

I hope this section title does not make anyone think that I plan to digress into some sort of weird science fiction novel, *The Mole That Ate Cleveland* or something. (My apologies to any readers from Cleveland. I have visited your city and found it to be very nice.) Consider the following reaction, which we will use again in the section on acid-base reactions.

$$HCl + NaOH \rightarrow NaCl + H_2O \qquad [Eq.\ 1\text{--}3]$$

This says that one molecule of hydrochloric acid will react with one molecule of sodium hydroxide to produce one molecule of table salt plus one molecule of water. Common sense says that no normal chemical reaction proceeds one molecule at a time. The term that was developed

CHAPTER 1: GENERAL CHEMISTRY

to deal with chemical matter and reactions in understandable quantities is known as the mole. The mole is defined as that amount corresponding to the atomic mass taken in grams. If we refer to Table 1–2 then, a mole of hydrogen atoms has a mass of 1.0078 grams. Similarly, a mole of oxygen has 15.9994 grams, a mole of iron 55.84 grams, and so forth. The same principle can be applied to molecules. Consider water, the primary subject of this book. The molecular weight (M.W.) is the sum of the individual atomic weights of oxygen (15.9994) and two hydrogen atoms (2 x 1.0078 = 2.0156) for a total of 18.015. One mole of water has a mass of 18.015 grams. Sodium chloride has a molecular weight of 58.4428, and a mole weighs 55.4428 grams.

So, how many atoms or molecules are in a mole? The Italian scientist Avogadro determined that a mole of any substance has 6.023×10^{23} (thank goodness for scientific notation) individual atoms or molecules as the case may be.

Using this concept, equation 1–3 can be restated to say that one mole of hydrochloric acid will react with one mole of sodium hydroxide to produce one mole of table salt plus one mole of water. Before leaving this idea, let's look at one other reaction.

$$Zn(OH)_2 + 2HCl \rightarrow ZnCl_2 + 2H_2O \qquad [Eq.\ 1\text{--}4]$$

In this reaction, one mole of zinc hydroxide reacts with two moles of hydrochloric acid to produce one mole of zinc chloride plus two moles of water. The masses of reactants and products are as follows:

$Zn(OH)_2$ M.W. = 99.3856   Mass reacted = 99.3946g

HCl   M.W. = 36.4609   Mass reacted = 2 x 36.4609g = 72.9218g

$ZnCl_2$ M.W. = 136.286   Mass reacted = 136.286g

H_2O M.W. = 18.0152   Mass reacted = 2 x 18.0152g = 36.0304g

We can see that this is correct, as the product mass exactly equals the mass of reactants (172.3164 grams).

One Other Concept

Before proceeding to reaction mechanisms, I want to introduce one other simple concept. Later sections and chapters will outline reactions first performed by scientists as experiments or to establish base-line data. Many of these were conducted under conditions known as "standard temperature and pressure," or STP for short. STP is defined as one atmosphere of pressure at 25°C (77°F).

Acid-Base Reactions

Acids and bases are among the most common chemicals on the planet and are manufactured and used by industry in enormous quantities. Even the power industry uses a reasonable share. The definition of an acid has evolved over the years along with the knowledge of chemistry. Very early descriptions of an acid included the following:

- Tastes sour
- Turns blue litmus dye red
- Reacts with carbonate minerals to produce carbon dioxide
- Neutralizes a base
- Reacts with metals and causes the evolution of hydrogen gas

All of these are true, and as we shall see elsewhere, the latter statement describes a corrosion reaction, but none are exceptionally scientific. The descriptions of a base were not any better.

- Tastes bitter
- Feels slippery
- Turns blue litmus red
- Absorbs carbon dioxide
- Neutralizes an acid

As chemistry became more precise, several better descriptions of acids and bases evolved. They include the following:

- An acid is a substance that produces the hydronium ion (H_3O^+) in water and a base produces the hydroxyl ion (OH^-) in water (Arrhenius theory).
- An acid is a proton (hydrogen ion) donor and a base is a hydrogen ion acceptor (Bronsted theory).
- An acid is an electron pair acceptor and a base is an electron pair donor (Lewis theory).

The Arrhenius theory only accounted for acid-base behavior in water, so the Bronsted and Lewis theories came about as scientists looked for ways to explain non-aqueous acid-base chemistry. For the purposes of this book the Arrhenius and Bronsted theories work well, so I will not discuss the Lewis theory.

The Arrhenius theory is illustrated by the following reactions:

$$HCl + H_2O \rightarrow H_3O^+ + Cl^- \qquad [Eq.\ 1–5]$$

$$NaOH + H_2O \rightarrow Na^+ + OH^- + H_2O \qquad [Eq.\ 1–6]$$

When hydrochloric acid is added to water it completely ionizes to produce hydronium ions. Sodium hydroxide produces the hydroxyl ion. If these two compounds are mixed directly, the Bronsted theory is applicable.

$$HCl + NaOH \rightarrow NaCl + H_2O \qquad [Eq.\ 1–7]$$

Hydrochloric acid donates a proton to sodium hydroxide, which accepts it. It is easy to see that if we mixed aqueous solutions of hydrochloric acid and sodium hydroxide, both the Arrhenius and Bronsted theories would have been validated. Personally, I would be reluctant to mix a strong acid and strong base, even in aqueous form, as the resulting reaction might shower me with corrosive chemicals. Examples of acid and base reactions will appear elsewhere in the text.

Addition and Elimination Reactions

An addition reaction occurs when two, possibly more, molecules combine to form a single molecule. This increases the number of bonds (coordination number) on an atom from each molecule. In the experiment I mentioned earlier, where I released a cloud of boron trifluoride, what I was actually attempting to do was perform an addition reaction, as outlined by the following equation:

$$(CH_3)_3N + BF_3 \rightarrow (CH_3)_3N-BF_3 \qquad [Eq.\ 1\text{–}8]$$

The experiment eventually turned out to be a success. This reaction produces so much heat that I had to conduct it in a glass vessel cooled with liquid nitrogen (-320°F). It's amazing how many pieces a banana will break into after being dipped in liquid nitrogen.

The reverse of an addition reaction is an elimination reaction. One molecule may split into two separate molecules. For example, calcium carbonate when heated breaks down via the following reaction:

$$CaCO_3 + heat \rightarrow CaO + CO_2 \uparrow \qquad [Eq.\ 1\text{–}9]$$

Substitution Reactions

Many chemical reactions involve a transfer or swap of molecules or atoms. For example, a common method to reduce calcium non-carbonate hardness in raw water is treatment with sodium carbonate.

$$CaSO_4 + Na_2CO_3 \rightarrow CaCO_3 \downarrow + Na_2SO_4 \qquad [Eq.\ 1\text{–}10]$$

Substitution of the carbonate ion for sulfate generates a product ($CaCO_3$) that is less soluble and precipitates from solution.

Oxidation-Reduction Reactions

The reactions outlined in the preceding sections involve a rearrangement of molecules. However, many reactions are known in which electrons move from one compound or element to another. Consider a simple experiment recommended in junior chemistry books. It is the copper plating of iron. In this experiment, an iron nail or other iron object is suspended in an aqueous solution of copper sulfate ($CuSO_4$). The iron atoms, all being equivalent, have a valence of zero. The copper atoms have a valence of +2 to counterbalance the sulfate valence of -2. Copper hangs on more tightly to its outer electrons than does iron, so in this experiment a lower energy state is reached when copper trades electrons with iron. The following reaction occurs.

$$CuSO_4 + Fe \rightarrow Cu + FeSO_4 \qquad [Eq.\ 1\text{--}11]$$

Written with valence or oxidation numbers, the reaction appears as:

$$Cu^{+2} + SO_4^{-2} + Fe^0 \rightarrow Cu^0 + Fe^{+2} + SO_4^{-2} \qquad [Eq.\ 1\text{--}12]$$

Metallic copper reverts to a zero oxidation state, so as this reaction proceeds copper plates out on the nail and iron atoms enter the solution to balance the electrical charge. Iron, in going from a zero to a +2 valence state is said to be oxidized, and copper is said to be reduced.

Electrochemical reactions are the basis upon which batteries operate, and it is also the chemistry that drives corrosion reactions. A very common corrosion cell, and one which is of extreme importance in steam generating systems, is shown in Figure 1–3. The illustration shows corrosion of a mild steel pipe wall, such as might be found in a cooling water or feedwater line. Here we see an electrochemical reaction in which iron oxidizes and goes into solution, while its valence electrons travel along the pipe wall to react with and reduce oxygen at another location. The location at which iron is oxidized is known as the anode and where oxygen is reduced, the cathode. Prerequisites for reactions of this type are these:

- An anode
- A cathode
- A path for electron flow (the pipe wall)
- An electrolyte

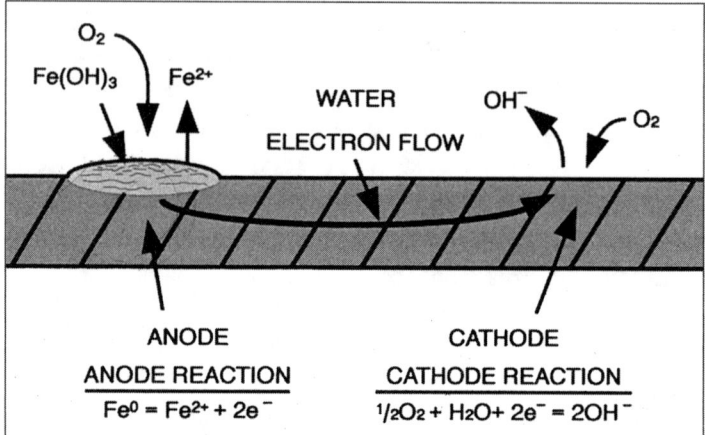

Fig. 1–3 *An Electrochemical Reaction*
Source: Betz Handbook of Industrial Water Conditioning, Ninth Edition. Betz Dearborn is a division of Hercules, Inc.

As we learned earlier, one of the early descriptions of an acid said that an acid reacts with metals and causes the evolution of a hydrogen gas. This is one of the primary corrosion mechanisms of metals like iron. An example reaction is shown below.

$$Fe + 2HCl \rightarrow FeCl_2 + H_2 \uparrow \qquad [Eq.\ 1\text{--}13]$$

In this reaction, iron transfers two electrons to the hydrogen nuclei, producing hydrogen gas. The equation can be re-written to show the electrochemical process.

$$Fe \rightarrow Fe^{+2} + 2e^- \qquad [Eq.\ 1\text{--}14]$$

$$2H^+ + 2e^- \rightarrow H_2 \qquad [Eq.\ 1\text{--}15]$$

Equations 1–14 and 1–15 are known as half-cell reactions. Any half-cell reaction has an electrochemical potential, which corresponds to the compound's ability to donate or accept electrons. Scientists have calculated electrical potentials for a large number of half-cell reactions. As a starting point, the half-cell reaction shown in Equation 1–15 is given a potential of zero. All other half-cell potentials derive from this standard. Table 1–3

CHAPTER 1: GENERAL CHEMISTRY

Oxidation-reduction Potentials

Half-cell Reaction	Oxidation-reduction Potential (volts)
$K \rightleftharpoons K^+ + e^-$	−2.92
$Na \rightleftharpoons Na^+ + e^-$	−2.71
$Al \rightleftharpoons Al^{+3} + 3e^-$	−1.66
$Fe \rightleftharpoons Fe^{+2} + 2e^-$	−0.44
$H_2 \rightleftharpoons 2H^+ + 2e^-$	0.00
$Cu \rightleftharpoons Cu^{+2} + 2e^-$	0.34
$Au \rightleftharpoons Au^{+3} + 3e^-$	1.42

Table 1-3 *Oxidation-reduction Potentials*

shows oxidation-reduction potentials for some of the most common elements or compounds.

Reactions having a negative potential tend to proceed from left to right, while those with a positive potential do not, but would rather proceed in the reverse direction. It is easy to understand why elements such as potassium, sodium, and aluminum have such large negative potentials. These atoms have outer electrons that are very loosely held. Conversely, noble metals like gold do not easily give up electrons. This is why natural deposits of gold exist in the elemental state. Oxidation-reduction potentials also help in explaining the iron-nail-in-copper-sulfate experiment mentioned above. Iron has a reasonably strong tendency to donate electrons, as is evidenced by an oxidation-reduction potential of −0.44V. Copper in the +2 state has a tendency to accept electrons, as the reverse of the reaction shown in the table has a potential of −0.34V. Together, the two half reactions produce an overall electromotive potential of −0.78V for the reaction.

I heard a story once of a photographic laboratory, in which the managers, after many years of being in business, decided that the plumbing waste lines needed to be replaced. The lab used copious amounts of silver nitrate [$Ag(NO_3)_2$] for film developing. Before I continue, look at the following half-cell potential for silver ions:

$$Ag \rightleftharpoons Ag^{+2} + 2e^- \qquad Eo = +0.80V \qquad [Eq.\ 1\text{–}16]$$

The plumber that they called in to replace the lines put in the contract that he be allowed to salvage the old material. He took the old pipes to a precious metal dealer and made a considerable sum. Similar to the example of copper plating above, the silver ions plated out on the iron-based drainpipes. I use this example to refute those people who like to make jokes about plumbers.

A number of corrosion mechanisms will appear in later chapters. The important point to keep in mind is that these are electrochemically driven, and relate to the ability of elements to donate or accept electrons.

Equilibrium Reactions

Many of the reactions we have so far examined proceed to completion. Thus, if I added hydrochloric acid to sodium hydroxide, the reaction would continue until either the HCl or NaOH disappeared first. Many reactions do not go to completion but proceed to a point where reactants and products exist in steady concentrations. These are known as equilibrium reactions. A common example used in many textbooks of an equilibrium reaction is that of hydrogen and iodine in a heated (400°C) atmosphere.

$$H_2 + I_2 \rightleftharpoons 2HI \qquad [Eq.\ 1\text{–}17]$$

The extent of this reaction has been determined by experiment. Using this knowledge, an equilibrium constant can be written for the reaction.

$$K_{eq} = \frac{[HI]^2}{[H_2] \cdot [I_2]} \qquad [Eq.\ 1\text{–}18]$$

This equation says that the quantity of reactants in moles divided by the quantity of products (again in moles) will equal a constant. The equilibrium constant for this reaction happens to be 64. Thus, if we placed a known quantity of hydrogen and iodine in a container and heated the compounds to 400°C, we could predict the amount of HI that would form and the amounts of H_2 and I_2 that would be consumed.

Note that the top term in Equation 1–18 is squared since the reaction produces two moles of hydrogen iodide. Whenever a reactant or product has a value greater than one for the mole ratio, that value becomes an exponent in the equilibrium equation. A case in point is the reaction of nitrogen to produce ammonia.

$$N_2 + 3H_2 \leftrightarrows 2NH_3 \qquad [Eq. \ 1\text{--}19]$$

The equilibrium constant is

$$K_{eq} = \frac{[NH_3]^2}{[N_2] \cdot [H_2]^3} \qquad [Eq. \ 1\text{--}20]$$

Somewhat different terminology is used for reactions in water. Consider the following reaction when chlorine is added to water.

$$Cl_2 + H_2O \leftrightarrows HOCl + HCl \qquad [Eq. \ 1\text{--}21]$$

The equilibrium constant may be written as:

$$K_{eq} = \frac{[HOCl] \cdot [HCl]}{[Cl_2] \cdot [H_2O]} \qquad [Eq. \ 1\text{--}22]$$

The brackets around the reactants and products indicate the aqueous concentration, in this case moles per liter.

Equilibrium constants for reactions in water can be simplified, and for this reaction the equilibrium constant becomes this:

$$K_{eq} = \frac{[HOCl] \cdot [HCl]}{[Cl_2]} \qquad [Eq. \ 1\text{--}22]$$

This is possible, because in most aqueous-based reactions the concentration of water far outweighs the concentration of other species. Thus, the water concentration can be assumed to be constant and the equilibrium reaction can be simplified to reflect just the important species.

Conclusion

This chapter outlined some important fundamentals of chemistry. The following chapters describe the chemistry of water and its reaction in steam generating systems. Water, and its impurities, behave according to many of the principles already outlined.

Learning Aide 1-1

SCIENTIFIC NOTATION

When dealing with scientific concepts, we often encounter very large and very small numbers. For example, the number of water molecules in a can of soda soft drink is approximately 12 followed by 24 zeros! Imagine writing this number several times in a report! Scientific notation, which is based on exponential notation using the base 10, makes the task of reporting large and small numbers much easier.

Basic mathematics defines the following:

$10^1 = 10$

$10^2 = 10 \cdot 10 = 100$

$10^3 = 10 \cdot 10 \cdot 10 = 1,000$ and so forth

This is known as exponential notation, with ten being the base number and 1, 2, 3, etc., being the exponents. Using this notation, the number 1,000,000 can be written as 1×10^6. The key to writing scientific notation for numbers larger than zero is to start at the decimal point, or where it would be, and count the number of spaces to the left until reaching the space between the final two numbers. Place a decimal point at this space, and then use the number of spaces counted as the exponent in the exponential notation. Examples:

$43,000 = 4\ 3\ 0\ 0\ 0 = 4.3 \times 10^4$

$560,000,000 = 5\ 6\ 0\ 0\ 0\ 0\ 0\ 0\ 0 = 5.6 \times 10^8$

$3284.79 = 3\ 2\ 8\ 4\ .79 = 3.28479 \times 10^3$

The reverse procedure is required for very small numbers, where the exponential notation follows the pattern:

$10^{-1} = 0.1$

$10^{-2} = 0.01$

$10^{-3} = 0.001$

Examples:

$0.2 = 2 \times 10^{-1}$

$0.0025 = 0.\ 0\ 0\ 2\ 5 = 2.5 \times 10^{-3}$

$0.0000567 = 0.\ 0\ 0\ 0\ 0\ 5\ 6\ 7 = 5.67 \times 10^{-5}$

THE CHEMISTRY OF NATURAL WATER SUPPLIES | 2

Water is transported from place to place on earth via a process known as the "Hydrologic Cycle." In simple terms, the hydrologic cycle is the evaporation of water from oceans, lakes and vegetation, and its subsequent return to earth through various forms of precipitation. Water vapor may be transported a short distance or many miles before atmospheric conditions cause condensation and precipitation. Along the way, the moisture picks up gases from the atmosphere, including pollutants, which change its chemistry appreciably.

Precipitation that falls on a landmass follows one or more paths. It may accumulate in a lake, reservoir, or pond and re-evaporate. Or, the water may flow into streams and rivers and return to the ocean. A major portion percolates through soil to collect in underground aquifers. Both surface and groundwater are common sources for makeup, cooling, and other applications in industrial plants. The wide variation in chemistry of water supplies makes water treatment a challenging task. Before examining these chemistries, we will first look at the unique properties of the water molecule itself, and learn why it behaves as it does.

Chemistry of Water

"H two O." Even schoolchildren know the chemical formula for water. Simple, right? Well, not quite. Fortunately for mankind, the seemingly simple water molecule actually has some interesting chemical properties that allow life to exist as we know it.

Two of the most important properties are bond angle of the hydrogen atoms to oxygen and the polarity of the water molecule. Regarding bond angle, first consider the molecular structure of another common, three-atom molecule, carbon dioxide (Figure 2–1). This molecule is linear with each oxygen atom 180° apart. Now, examine the water molecule shown in Figure 2–2. Note that the molecule is not linear, but that hydrogen atoms bond at an angle of 104.5°. This geometrical arrangement is important, especially when taken in conjunction with the polarity of water molecules.

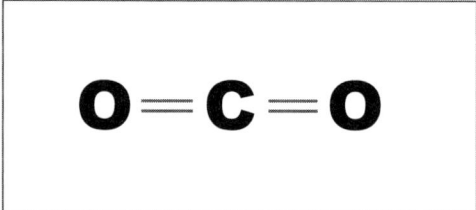

Fig. 2–1   *Carbon Dioxide Molecule*

As you may recall from chapter 1, oxygen is very electronegative with hydrogen being rather less so. Although the hydrogen-oxygen bonds in a water molecule are considered to be covalent, the highly electronegative oxygen atom exerts a strong influence on the hydrogen electrons, pulling them some distance away from the hydrogen nuclei. This imparts a partial negative charge to oxygen and a partial positive charge to hydrogen. Molecules of water resemble tiny magnets, which causes molecules to be electrostatically attracted to their neighbors. This phenomenon is known as hydrogen bonding. The strength of the hydrogen bond is only about one tenth that of the covalent bonds, but without hydrogen bonding life

CHAPTER 2: THE CHEMISTRY OF NATURAL WATER SUPPLIES

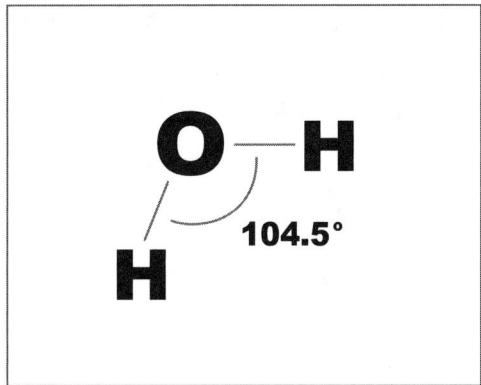

Fig. 2–2   *Water Molecule*

as we know it would not be possible. Scientists have calculated that the boiling point of water in the absence of hydrogen bonding would be around –100°C (–148°F) instead of the +100°C (212°F) temperature that we know.

The effects of hydrogen bonding are visually evident in the formation of ice. As water freezes, hydrogen bonds cause the molecules to form a crystalline structure that has a lower density than the liquid. Thus, ice floats. This is extremely important to nature, as the winter-time formation of an ice layer on ponds and lakes insulates the water below. Without hydrogen bonding, bodies of water in northern climates would freeze solid during winter.

The polar nature of the water molecule gives it excellent solvating properties towards other polar compounds. When water dissolves an ionic compound such as sodium chloride (NaCl), the water molecules actually surround the individual sodium (Na^+) and chloride (Cl^-) ions. This same property, however, makes water a poor solvent for non-polar compounds. A practical example of this can be seen around automobiles during a rainstorm, where any spilled oil visibly floats on puddles. Oil, being non-polar, is not attracted to water molecules and vice versa.

The ability of water to dissolve minerals is dependent upon a number of factors, including the bond strength of the mineral crystal lattice, mineral composition, temperature, and often the acidity or basicity of the water. More about this later.

Self-Ionization of Water, Acidity, Basicity, and pH

Another very important property of water is self ionization. In pure water, a small fraction of the molecules dissociates into hydrogen ions and hydroxyl ions.

$$H_2O \rightleftharpoons H^+ + OH^- \qquad [Eq.\ 2\text{--}1]$$

Hydrogen ions do not exist by themselves in solution, and Equation 2–1 is better represented by

$$2H_2O \rightleftharpoons H_3O^+ + OH^- \qquad [Eq.\ 2\text{--}2]$$

Experiment has shown that in a pure solution the concentration of hydrogen ions and hydroxyl ions is 1×10^{-7} moles per liter. By convention, the equilibrium constant for this reaction is given as

$$K_e = K_w = [H_3O^+] \cdot [OH^-] = (1 \times 10^{-7}) \cdot (1 \times 10^{-7}) = 1 \times 10^{-14} \qquad [Eq.\ 2\text{--}3]$$

Any increase in H^+ ions will decrease the OH^- concentration and vice versa. This leads us to the subject of acids and bases and the pH scale.

In chapter 1 we learned that one definition of an acid is a substance that adds hydrogen ions to water. A very well known acid is hydrochloric, which adds hydrogen ions via the following reaction.

$$HCl + H_2O \rightarrow H_3O^+ + Cl^- \qquad [Eq.\ 2\text{--}4]$$

A base has the opposite effect and increases the hydroxyl ion concentration. Sodium hydroxide is the most well known base.

$$NaOH + H_2O \rightarrow OH^- + H_2O + Na^+ \qquad [Eq.\ 2\text{--}5]$$

How do we measure acidity and basicity? We know that a pure solution of water contains a 1×10^{-7} moles per liter concentration of both hydrogen ions and hydroxyl ions. Let's say we added enough hydrochloric acid to increase the hydrogen ion concentration to 1×10^{-6} moles per liter. (And, since $[OH^-] = K_w/[H^+]$, decrease the hydroxyl ion concentration to 1 x 10^{-8} moles/l.) We could say the solution has an acidity of 1 x 10^{-6} moles per liter. This is a tedious method to express one's self. Even though science sometimes seems complicated, scientists are always looking for ways to make things easy. The pH scale represents a simplification of the terminology for expressing acid and base concentrations. The pH of a solution is the negative logarithm of the hydrogen ion concentration in moles per liter.

Although the electronic calculator has made the calculation of logarithms very simple, let's briefly look at them the old-fashioned way. Every number can be expressed as a power of ten. For example, 100 is equal to 10 x 10 or 10^2. Ten is the base number, and two is the power or logarithm. The number 1000 can be represented as 10^3, where three is the logarithm. For numbers that are not exact multiples of 10, the logarithms have fractional components. The number 50 is $10^{1.6989}$ in logarithmic terms. The same principle applies to numbers smaller than one. For our example above, where the hydrogen ion concentration is 1 x 10^{-6} (0.000001) moles/l, the logarithm is -6. The negative logarithm is 6, which by definition is the pH of the solution. A solution with a hydrogen ion concentration of 1 x 10^{-5} moles per liter has a pH of 5, and so forth. Each change in pH unit represents a ten-fold change in hydrogen and hydroxyl ion concentrations.

A pH of 7 represents neutrality. Any increase in pH above 7 correlates to an increase in hydroxyl ions and a corresponding increase in the basicity of the solution. So, if we added some sodium hydroxide to increase the hydroxyl ion concentration to 1 x 10^{-6} moles per liter, the hydrogen ion concentration would, of course, be reduced to 1 x 10^{-8} moles per liter. This corresponds to a pH of 8. Table 2–1 illustrates the hydrogen and hydroxyl ion concentration over the normal pH range.

The acidity or basicity of a water solution can greatly affect its reactivity towards many compounds. Examples of this appear in the following sections, and as we shall see in later chapters, the pH of steam generation fluids must be maintained within a moderately alkaline range to minimize corrosion.

In the examples of an acid and base above, I chose hydrochloric acid and sodium hydroxide because these are very common materials. Each completely ionizes in water. Many acids and bases, including quite a few

Relationship of pH to Hydrogen and Hydroxyl Ion Concentrations

H+	OH+	pH
10^{-1}	10^{-13}	1
10^{-2}	10^{-12}	2
10^{-3}	10^{-11}	3
10^{-4}	10^{-10}	4
10^{-5}	10^{-9}	5
10^{-6}	10^{-8}	6
10^{-7}	10^{-7}	7
10^{-8}	10^{-6}	8
10^{-9}	10^{-5}	9
10^{-10}	10^{-4}	10
10^{-11}	10^{-3}	11
10^{-12}	10^{-2}	12
10^{-13}	10^{-1}	13

Table 2–1 *Relationships: pH to Hydrogen ion concentration*

found in nature, do not completely ionize, but rather reach an equilibrium. Consider acetic acid. When placed in water the reaction is:

$$HC_2H_3O_2 + H_2O \leftrightharpoons H_3O^+ + C_2H_3O_2^- \qquad [Eq.\ 2\text{--}6]$$

The equilibrium equation for this reaction is:

$$K_{eq} = \frac{[H_3O^+] \cdot [C_2H_3O_2^-]}{[HC_2H_3O_2]} \qquad [Eq.\ 2\text{--}7]$$

At standard conditions, K_{eq} equals 1.8×10^{-5}. Thus, this acid only slightly ionizes in water. Acids of this type are known as weak acids, and the smaller the value of K_{eq}, the weaker the acid. Bases may follow a similar pattern. Examine the reaction of trimethyl amine with water:

$$(CH_3)_3N + H_2O \rightleftharpoons (CH_3)_3N\text{--}H^+ + OH^- \qquad [Eq.\ 2\text{--}8]$$

The equilibrium equation for this reaction is:

$$K_{eq} = \frac{[(CH_3)_3NH^+] \cdot [OH^-]}{[(CH_3)_3N]} \qquad [Eq.\ 2\text{--}9]$$

Similar, to the weak acid example above, trimethylamine only slightly ionizes, so it is considered a weak base.

Changes in chemical or physical conditions will affect equilibrium reactions, and may cause an increase in reactant or product concentration. This process is known as Le Châtelier's Principle.

Le Châtelier's Principle

Refer again to the equilibrium reaction shown in Equation 2–6. What would happen if we added some acetic acid to the solution? The reaction would immediately shift to the right, and more $HC_2H_3O_2$ would ionize. Le Châtelier's Principle says that if a change is made to the properties of a system, the system will change in an effort to re-establish equilibrium. Had we added sodium acetate ($NaC_2H_3O_2$) to the solution, and increased the concentration of $C_2H_3O_2^-$ ions, the reaction would have shifted to the left.

Not only chemical but physical factors can affect systems at equilibrium. Temperature is one of these factors, as we shall see. Le Châtelier's Principle is important with regard to water chemistry, and several practical examples appear throughout this book.

Di- and Triprotic Acids

Hydrochloric and acetic acids, our two examples from above, are monoprotic acids. That is, each only donates one hydrogen ion (proton) in a chemical reaction. Other acids exist, however, that can donate more than one hydrogen ion. Two of the most common are sulfuric acid (H_2SO_4) and phosphoric acid (H_3PO_4). They represent di- and triprotic acids, respectively. Sulfuric acid is a strong acid, and donates the first of its two hydrogen ions quite readily. The second is held a bit more tightly. Phosphoric acid is much weaker, as is evidenced by the ionization constants of the respective acids:

$$H_3PO_4 + H_2O \rightleftharpoons H_2PO_4^- + H_3O^+ \quad K_{ion} = 7.5 \times 10^{-3} \quad [Eq.\ 2\text{--}10]$$

$$H_2PO_4^- + H_2O \rightleftharpoons HPO_4^{-2} + H_3O^+ \quad K_{ion} = 6.2 \times 10^{-8} \quad [Eq.\ 2\text{--}11]$$

$$HPO_4^{-2} + H_2O \rightleftharpoons PO_4^{-3} + H_3O^+ \quad K_{ion} = 4.8 \times 10^{-13} \quad [Eq.\ 2\text{--}12]$$

Each succeeding hydrogen ion is more difficult to remove. I specifically used phosphoric acid and its species for this example, because the phosphate ion (PO_4^{-3}) has been a primary treatment chemical in boilers for years. The phosphate species act as buffering agents in boiler water.

Buffers

Buffers are compounds that will absorb excess acid or alkalis that may be introduced to the solution. As Equations 2–10 through 2–12 illustrate, the phosphate ion can exist in several different forms. Each of the reactions is an equilibrium reaction and, depending on conditions, the phosphate species can either accept or donate hydrogen ions. If excess acid or alkali enters the system, the chemistry adjusts to maintain equilibrium. Let's look again at Equation 2–12. The K_{eq} is very small at 4.8 x 10^{-13}. If we turned the reaction around, the new equilibrium constant would be $1/K_{eq}$ or 1/4.8 x 10^{-13} = 2.1 x 10^{12}. This says that phosphate ion added to an acid solution would immediately reduce the acidity by absorbing hydrogen ions. Conversely, excess hydroxide ions would be neutralized by accepting hydrogen ions from protonated phosphate compounds. This principle is the basis behind one of the most common boiler water treatment programs, which we will learn more about in the chapter on internal water treatment.

Buffers are found throughout nature and are very important. For instance, human blood is buffered at a pH of 7.14. Any change can produce sickness or worse. As a college text of mine stated, "Some lapse of judgement could have put you to the discomfort...that results from overexposure to dilute and sweetened ethanol. This involves a small and hopefully temporary displacement of the blood pH." At the time when I first read this text, I was at an age when I occasionally tested this "theory." The authors were correct. A shift in blood pH does cause discomfort. If any of you readers ever contact me, ask about the game of golf I tried to play after one late night college party. Jack Nicklaus I was not.

One of the most important natural buffering systems is the carbon dioxide, bicarbonate, carbonate complex. Figure 2–3 shows the distribution of these species as a function of pH. Carbon dioxide is continually being transferred from the atmosphere to earth and back, but CO_2 is also produced by microbiological action in soil. Rainwater actually picks up

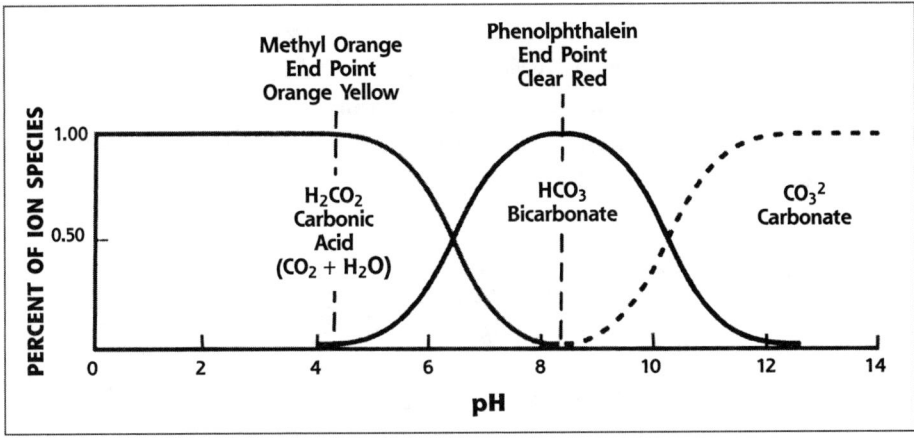

Fig. 2–3   *Relationship of CO_2, HCO_3^-, and CO_3^{-2} in water.*

much more CO_2 in the upper layers of the soil than from the atmosphere. Much of this CO_2 is converted into bicarbonate and smaller amounts of carbonate through reactions with soil compounds. The diagram shows that the acidity or basicity of the water determines the ratio of the various species. Note that carbon dioxide and carbonate cannot exist together in solution. A similar situation occurs with the phosphate species listed earlier. The most acidic and basic species cannot co-exist in the same solution.

The Chemistry of Natural Waters

As we started to learn in the preceding section, natural waters contain many impurities. We will examine in more detail the processes by which water absorbs impurities.

Effects of the Atmosphere on Water Chemistry

Table 2–2 lists the major components of the earth's atmosphere. While nitrogen and oxygen by far make up the bulk of atmospheric compounds, other chemicals exist in significant quantities. Most important is carbon dioxide. As we have seen, moisture picks up CO_2 from the atmosphere and from the soil. Carbon dioxide reacts with water to produce carbonic acid, which by itself lowers pH.

Major Constituents of the Lower Atmosphere

Component	Percent by Weight
Nitrogen	75.53%
Oxygen	23.14
Argon	1.28
Carbon Dioxide	0.05
Neon	1.25×10^{-3}
Krypton	2.89×10^{-4}
Ozone	1×10^{-4}
Helium	7.2×10^{-5}
Xenon	3.6×10^{-5}
Hydrogen	3.48×10^{-6}

* *Dry atmosphere.*

Table 2–2 *Major Components of the Earth's Atmosphere*

CHAPTER 2: THE CHEMISTRY OF NATURAL WATER SUPPLIES

$$CO_2 + H_2O \leftrightharpoons H_2CO_3 \qquad [Eq.\ 2\text{-}13]$$

Natural rainwater has a pH of around 5.7. Industrial processes add pollutants to the atmosphere, which may lower the pH even further. Power plants and industrial boilers introduce sulfur dioxide (SO_2) and oxides of nitrogen (NO_x), and automobiles contribute to the NO_x pollution. These compounds oxidize further and eventually combine with moisture to produce sulfuric acid and nitric acid (HNO_3), the two primary ingredients of acid rain. Readings as low as a pH of 3.8 have been recorded in areas prone to acid precipitation.

Precipitation, whether it be tainted or untainted by manmade acids, certainly is active. It is now time to examine the compounds and minerals that water gathers as it travels along or through the earth's crust.

Common Elements and Compounds in the Earth's Crust

According to spectrographic observations and calculations by astronomers, the matter in the universe is composed of 75% hydrogen with helium making up virtually all of the balance. Yet the earth obviously contains many more elements than this, and in much greater ratios. How is this possible? The answer lies in the stars. Most of the energy produced by stars comes from the thermonuclear reaction of hydrogen to form helium. This is the predominant process in an average star such as our sun, whose core temperature is estimated to be 45,000,000°F or so. In very large stars, which may have core temperatures of 5.4 billion degrees Fahrenheit or more, additional nuclear reactions occur, and heavier elements such as carbon and oxygen form. Still, this does not account for even heavier elements. In all stars the supply of hydrogen will eventually dwindle and the star will begin to die. What happens next is fascinating. As the hydrogen-to-helium reaction falls off, the heat produced no longer counterbalances the enormous force of gravity generated by the star's mass and the core begins to collapse. The core then heats causing additional nuclear reactions. In stars such as the sun, this process is not very violent, and often only causes the outer layers of the star to expand. (Even so, when this happens with the sun, life on earth will perish, as the sun's outer layers may expand to the orbit of Venus and perhaps beyond.)

In more massive stars the gravitational force is greater and the death process becomes more violent. Very large stars explode, producing the phenomena known as supernova. A supernova produces so much energy that many additional nuclear reactions occur, generating elements up to uranium and perhaps beyond. (The death of a large star also may produce a black hole, but that is a different story.) Astronomers believe that the earth and other planets in the solar system are remnants of a supernova.

Supernova remnant or not, the earth has an interesting makeup. Table 2–3 shows the most common elements in the earth's crust. The most abundant elements are oxygen, silicon, aluminum, iron, and several of the alkaline earths and alkalis. When moisture reaches the earth's surface it reacts with and dissolves minerals and compounds in the soil. The path taken by water greatly influences its subsequent composition. Let's exam-

Composition of the Earth's Crust

Component	Percent by Weight	Component	Percent by Weight
Oxygen	46.43%	Sulfur	0.05%
Silicon	27.77	Barium	0.05
Aluminum	8.13	Chromium	0.04
Iron	5.12	Zirconium	0.03
Calcium	3.63	Carbon	0.03
Sodium	2.85	Vanadium	0.02
Potassium	2.60	Nickel	0.02
Magnesium	2.09	Strontium	0.02
Titanium	0.63	Lithium	0.003
Phosphorus	0.13	Copper	0.002
Hydrogen	0.13	Cerium	0.0015
Manganese	0.10	Beryllium	0.001
Fluorine	0.08	Cobalt	0.001
Chlorine	0.06	All Others	0.0078

Table 2–3 *Most Common Elements in the Earth's Crust*
Source: Betz Handbook of Industrial Water Conditioning, Ninth Edition. Betz Dearborn is a division of Hercules, Inc.

ine groundwater first. As water passes through soil to collect in underground aquifers, any number of reactions may occur. As we have seen, soil is principally composed of silicates, with other ions mixed in. Even though silicates are not highly soluble, enough will dissolve to generate a silica (SiO_2) concentration of several parts-per-million (ppm) or so. Some underground supplies, especially those located beneath sandy soil, may contain more than 50 ppm of silica. Silica is an extremely deleterious compound in steam generating systems, and it, or silicate compounds, will form deposits on turbine blades, superheater components, and boiler tubes. Groundwaters also pick up calcium and magnesium, the hardness ions. These form scale in steam generating, cooling water, and makeup treatment systems. The concentrations of calcium and magnesium may become quite high if the water passes through limestone beds or collects in limestone aquifers. Water's reactivity towards limestone is enhanced by acidity, which of course is naturally produced by absorption of carbon dioxide. The representative reactions when water passes through limestone deposits are

$$CaCO_3 + H_2CO_{3\ (aq)} \rightleftharpoons Ca(HCO_3)_2 \qquad [Eq.\ 2\text{--}14]$$

$$MgCO_3 + H_2CO_{3\ (aq)} \rightleftharpoons Mg(HCO_3)_2 \qquad [Eq.\ 2\text{--}15]$$

Not only do these reactions increase the hardness of the water, but the bicarbonate alkalinity goes up as well. Sodium and, to a lesser extent, potassium make up many of the dissolved cations in water. Although it is not as obvious, chloride and sulfate often make up the balance of anions in groundwater supplies. Iron and manganese are generally present in low concentrations of a few ppm or less. Even so, they can be very troublesome. Both usually exist in a non-oxidized form since groundwater is not exposed to the atmosphere. Once the water comes in contact with air, iron and manganese react with oxygen to produce dark colored deposits that stain plumbing fixtures and clothes, and in industry foul makeup treatment equipment including filters and ion exchange units.

Ground waters will also absorb gases trapped underground, the most notorious of which is hydrogen sulfide (H_2S). As water percolates through soil, the soil acts as a filtering medium. So, although groundwaters usually contain large quantities of dissolved solids, suspended solids concentrations are very low. The reverse is usually true with surface waters. Dissolved solids concentrations are lower, but suspended solids in the form of mud, silt, organic matter, colloidal silica, microorganisms, and even general debris such as leaves and trash are greater. We will examine these foulants in detail in subsequent chapters. Mud and silt may be particularly troublesome in cooling water systems. Organics, produced by decaying vegetation, can foul makeup treatment systems, most notably ion exchange resins. Colloidal silica exists as very small particles (colloids) that pass through ion exchange beds and then revert to ionic silica when exposed to heat in a boiler.

Surface waters also accumulate agricultural chemicals from runoff. At the electric and water utility where I formerly worked, makeup and cooling water come from a lake. We would see the nitrate, ammonia, and organic carbon concentrations increase every spring when rains washed farm chemicals into the lake. Water treatment personnel always have to

Chemistry of Selected Water Supplies
(all values except pH expressed as ppm)

Constituent	Well Water near Tecumseh, KS	Surface Water near Beaumont, TX	Tombigbee River Alabama	Sea Water
Calcium	120.0	8.7	16.2	400
Magnesium	23.5	2.8	4.1	1,272
Potassium	6.6	2.6	2.0	380
Sodium	73.4	14.8	15.3	10,561
Bicarbonate	297.6	34.6	54.3	
Carbonate	0.0	0.0	0.0	
Chloride	99.3	19.0	16.4	18,980
Nitrate	0.0	0.0	1.5	
Phosphate	0.0	0.0		
Sulfate	150.0	17.0	30.2	2,652
Iron	0.4	2.3	3.6	
Manganese		0.1	0.3	
Silica (SiO_2)	19.0	12.4	3.1	8.3 max.
Total Organic Carbon	0.0	6.7		
pH	7.7	7.7	7.2	8

Table 2–4 *Chemistry of Selected Water Supplies*

increase the activated carbon feed at this time of year to absorb the organic pesticides and herbicides that enter with the spring rains.

Table 2-4 illustrates the chemistry of several water supplies in the United States. Some of the trends we just mentioned are evident. The groundwater supply has a much higher concentration of virtually all of the major ions. Surprisingly, iron is higher in the surface supplies. Often, the reverse is true. Note that the surface supply near Beaumont, Texas shows a significant concentration of organic carbon. This area of the country is very marshy, and in fact, rice is grown there. Small wonder that organic carbon should be present in the supply. Much variability also occurs within individual supplies, and especially surface sources. Table 2-5 shows the maximum and minimum values of several impurities in sam-

Chemistry of Missouri River Water at Kansas City
1994 Maximum and Minimum Values
(all values except pH measured in mg/l)

Constituent	Minimum Concentration	Maximum Concentration
Calcium	69.7	93.0
Magnesium	25.6	31.3
Sodium*	37.6	64.8
Bicarbonate	197.5	293.8
Chloride	23.0	30.0
Sulfate	131.0	182.0
Silica	8.5	57.7
Total Solids	550.0	1708
pH	7.9	8.4

* *Sodium and potassium grouped together.*

Table 2-5 *Chemistry of the Missouri River at Kansas City*

ples taken from the Missouri River in Kansas City. Notice the very large difference between minimum and maximum suspended solids. Silica concentrations are also highly variable.

The impurities in natural water supplies would cause great damage if allowed to enter a steam generating system, and we will examine pretreatment methods in the next chapter. Before ending this chapter though, let's look at the units by which dissolved and suspended solids are measured in water supplies and also examine a concept known as the solubility product.

Units of Measurement

Concentrations of dissolved or suspended solids in natural water supplies are usually measured in parts-per-million (ppm) or milligrams per liter. As the name implies, a part-per-million says that for every one part (by weight) of dissolved or suspended solid there are one million parts of water. Because a part-per-million almost, but not quite exactly, equals a milligram per liter (mg/l), the units are often interchanged. For very pure water, such as that produced by sophisticated makeup treatment systems, concentrations are often measured in parts-per-billion (ppb) or the near equivalent, micrograms per liter (µg/l). New steam guidelines for combined-cycle generating units are even mentioning a few contaminants in the part-per-trillion range.

When examining concentrations or reactions of ions in water, a person must make sure that the units of measurement are consistent. For example, in a solution containing calcium ions (Ca^{+2}) and chloride ions (Cl^-), each calcium will pair with two chlorides. So, one mole of calcium, weighing 100 grams will pair with two moles of chloride weighing 70.9 grams. Replace the chlorides with sulfate ions, (SO_4^{-2}), and one calcium pairs with one sulfate. (100 grams with 96 grams.) One must keep track of equivalent units to avoid mistakes when calculating the weight of elements or compounds in solution. A term developed to account for this difference is known as equivalent-per-million (epm). The epm may be calculated by dividing a substance's molecular weight by its normal valance. Personally, I find the epm method difficult to work with. A simpler method is that based on calcium carbonate as a standard. Calcium carbonate has a molecular weight very close to 100. Scientists have developed conversion factors that allow one to calculate the weights of most water constituents in terms of parts-per-million as $CaCO_3$. This

$CaCO_3$ Equivalency Table for Common Ions
(all values except pH expressed as ppm)

Element or Compound	Chemical Formula	Molecular Weight	Multiplication Factors Substance to $CaCO_3$ Equivalent	$CaCO_3$ Equivalent to Substance
Aluminum	Al^{+3}	27.0	5.56	0.18
Ammonia	NH_3	17.0	2.94	0.34
Ammonium Ion	NH_4^+	18.0	2.78	0.36
Barium	Ba^{+2}	137.3	0.73	1.37
Calcium	Ca^{+2}	40.1	2.50	0.40
Carbon Dioxide	CO_2	44.0	1.14	0.88
Iron (Ferric)	Fe^{+3}	55.8	2.69	0.37
Iron (Ferrous)	Fe^{+2}	55.8	1.79	0.56
Hydrogen	H^+	1.0	50.00	0.02
Magnesium	Mg^{+2}	24.3	4.12	0.24
Manganese (Manganic)	Mn^{+3}	54.9	2.73	0.37
Manganese (Manganous)	Mn^{+2}	54.9	1.82	0.55
Potassium	K^+	39.1	1.28	0.78
Silica	SiO_2	60.1	0.83	1.20
Sodium	Na^+	23.0	2.18	0.46
Acid Radicals				
Bicarbonate	HCO_3^-	60.1	0.82	1.22
Carbonate	CO_3^{-2}	60.0	1.67	0.60
Chloride	Cl^-	35.5	1.41	0.71
Hydroxide	OH^-	17.0	2.94	0.34
Nitrate	NO_3^-	62.0	0.81	1.24
Phosphate	PO_4^{-3}	95.0	1.58	0.63
Sulfate	SO_4^{-2}	96.1	1.04	0.96

Table 2–6 *$CaCO_3$ Equivalency Table for Common Ions*

makes comparisons easy. Some of the most common conversion factors are shown in Table 2–6. Let's see how this calculation method works. In a natural environment, waters are electrically neutral. The charge of the dissolved cations should exactly balance the charge of dissolved anions. A simple method to determine if an analytical test of a water sample is valid is to compare the sum of the cations as $CaCO_3$ equivalents to the sum of the anions in the same terms. The two sums should agree within 10%. Table 2–7 shows an individual analysis taken from Table 2–4. The sum of the weights of the cations and anions do not match. Now look at the calcium carbonate equivalent concentrations. The ionic balance is well within recommended guidelines. Sources such as the U.S. Filter Water and Waste Treatment Data Book provide equivalents for a wide variety of elements and compounds.

Water Analysis
Ions and Ions as Calcium Carbonate

Cations	Concentration (ppm)	Concentration (ppm as CaCO$_3$)	Anions	Concentration (ppm)	Concentration (ppm as CaCO$_3$)
Calcium	120.0	300.0	Bicarbonate	297.6	244.0
Magnesium	23.5	96.8	Carbonate	0.0	0.0
Potassium	6.6	8.4	Chloride	99.3	140.0
Sodium	73.4	160.0	Nitrate	0.0	0.0
			Sulfate	150.0	156.0
TOTALS	223.5	565.2		546.9	540.0

Table 2–7 *Water Analysis: Ions as Calcium Carbonate*

Solubility of Minerals

Minerals and other compounds that dissolve in water will only do so to a certain extent, at which point the water becomes saturated with the compound. Any further addition will cause precipitation of the excess material. In effect, an equilibrium reaction occurs. The solubility of ionic solids and other materials varies greatly over the range of compounds. Factors that influence solubility include lattice strength of the mineral and the strength of electrostatic attraction between water molecules and the ions. These factors are in turn influenced by the charge on the ions, how well the ions fit together in the crystal lattice, and ionic size. Complex ions like the carbonates and sulfates are much larger than simple anions and cations, and their charge is spread out over a greater area. Temperature also has an effect on solvating properties, but not always in a manner you might expect.

CHAPTER 2: THE CHEMISTRY OF NATURAL WATER SUPPLIES

Solubilities of Mineral salts
(Grams per 100 ml)*

C/A	Fluoride	Chloride	Bromide	Iodide	Carbonate	Sulfate	Nitrate
Lithium	0.27	45.4	143	**	1.33	26.1	52.2
Sodium	4.22	36.1	94.6	184	29.1	28.0	91.0
Potassium	92.3	35.5	67.7	148	112	12.4	37.3
Rubidium	131	91.2	98.0	152	450	42.4	34.8
Cesium	**	186	124	44.0	261	167	14.9
Beryllium		**	**		0.36	42.5	**
Magnesium	0.0076	56.7	103	140	0.0716	36.4	72.7
Calcium	0.0017	74.5	153	208	0.000692***	0.066***	138
Strontium	0.011	55.8	107	178	0.00109	0.013	82.0
Barium	0.17	37.0	106	221	0.0024	0.00026	10.2

* Measurements taken at temperatures ranging from 15° to 25°C.
** No data available, but very soluble.
*** Data modified from original version. New data based on current solubility product constants.

Table 2-8 *Solubilities of Mineral Salts*

Table 2–8 shows the solubility of a number of ionic solids. Although there are a few anomalies in the data, the table clearly reveals many of the properties we examined in chapter 1. The increasing solubility (moving from left to right in the table) of the simple salts of the small alkalis (lithium and sodium), shows the instability that develops as anion size increases over that of the cations. The heavier alkali salts are also highly soluble because they consist of large cations and anions, where the charge is not as greatly centralized as in smaller ions. The opposite of this is seen with many of the fluoride salts, where the small fluoride ion generates a strong electrostatic field. The effect of multivalency on solubility is evident in the low solubilities of many of the alkaline earth-carbonate and sulfate salts. Magnesium sulfate, the ingredient in Epsom salt, is a notable exception, perhaps because the small magnesium cation does not fit in a lattice well with the large sulfate ion. The difference in solubility of calcium sulfate versus calcium carbonate is used to great effect in makeup water pretreatment, as we shall see later. The extremely slight solubility of barium sulfate can cause problems in a reverse osmosis system, which progressively concentrates impurities during the water purification process.

Temperature also affects the solubility of minerals, but this can work both ways. Higher water temperatures increase the solubility of many ionic salts, such as sodium chloride. For other salts, the opposite effect occurs. Unfortunately, this reverse solubility is most pronounced in several of the scale-forming salts. Figure 2–4 shows the change in solubility for two of the most common scale-forming materials, gypsum ($C_aSO_4 \cdot C_2H_2O$) and calcium carbonate. This effect explains why scale so often forms in heat exchangers. Chapters 4 and 7 will discuss scale formation on a practical level.

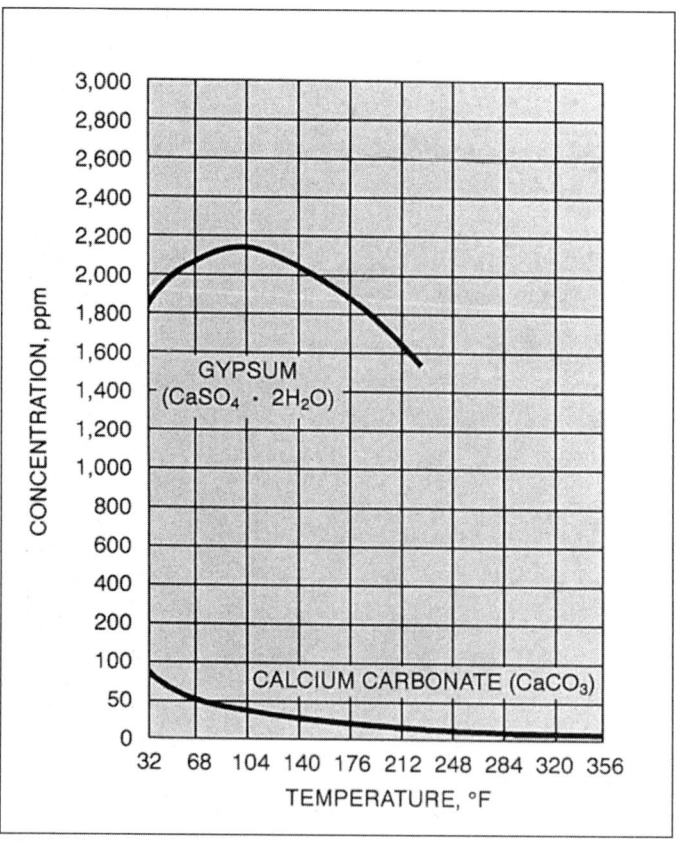

Fig. 2–4 *Solubility of Gypsum and Calcium Carbonate vs. Temperature* Source: *Betz Handbook of Industrial Water Conditioning, Ninth Edition. Betz Dearborn is a division of Hercules, Inc.*

CHAPTER 2: THE CHEMISTRY OF NATURAL WATER SUPPLIES

The Solubility Product

Similar to equilibrium reactions, the solubility of minerals can be described by an algebraic equation. Consider the dissolution of barium sulfate (a problematic RO foulant) in water. In a neutral water, only a small fraction of barium sulfate goes in to solution. The reaction is:

$$BaSO_4 \rightleftharpoons Ba^{+2} + SO_4^{-2} \qquad [Eq.\ 2\text{--}16]$$

We could write an equilibrium equation for this reaction:

$$K_{eq} = \frac{[Ba^{+2}] \cdot [SO_4^{-2}]}{[BaSO_4]} \qquad [Eq.\ 2\text{--}17]$$

However, this is not a true equilibrium reaction, since it is not solid barium sulfate that is breaking down to ions, but rather a dissolution process. We could put 1 gram or 1000 grams into water, but the amount that dissolves will still remain the same. This allows the equation to be simplified to the following:

$$K_{sp} = [Ba^{+2}] \cdot [SO_4^{-2}] \qquad [Eq.\ 2\text{--}18]$$

K_{sp} is known as the solubility product constant. These have been calculated for a wide variety of materials, and Table 2–9 shows some of the most common K_{sp} values. Let's see how to calculate the dissolution of barium sulfate, where K_{sp} equals 1.1×10^{-10}.

Solubility Product Constants at STP for a few Common Minerals

Al(OH)$_3$	Ksp = 1.9×10^{-33}
BaSO$_4$	Ksp = 1.1×10^{-10}
CaCO$_3$	Ksp = 4.8×10^{-9}
CaSO$_4$	Ksp = 2.4×10^{-5}
Fe(OH)$_3$	Ksp = 1.1×10^{-36}

Table 2–9 *Solubility Product Constants for Common Minerals*

- $K_{sp} = [Ba^{+2}] \cdot [SO_4^{-2}] = 1.1 \times 10^{-10}$
- Since Ba^{+2} and SO_4^{-2} are both equal, let each be represented as "x"
- $x^2 = 1.1 \times 10^{-10}$
- $x = $ Square root of $1.1 \times 10^{-10} = 1.05 \times 10^{-5}$

So, a saturated solution contains 1.05×10^{-5} moles per liter of barium ions and 1.05×10^{-5} moles per liter of sulfate ions. Since the molecular weight of barium is 137.3 grams and that of sulfate is 96.1 grams, it follows that

- The weight of barium in solution is 1.05×10^{-5} moles per liter $\cdot$ 137.3 grams per mole, which is 1.44×10^{-3} grams.
- The weight of sulfate in solution is 1.05×10^{-5} moles per liter $\cdot$ 96.1 grams per mole, which equals 1.01×10^{-3} grams.
- Adding the results together, the total weight of barium and sulfate in one liter of solution is 2.45×10^{-3} grams.

Let's look at another problem to get a better idea of how the solubility constants were set up by scientists. Consider another barium compound, barium fluoride (BaF_2). It dissolves in water according to the following reaction:

$$BaF_2 \rightleftharpoons Ba^{+2} + 2F^- \qquad [Eq.\ 2\text{--}19]$$

Equations

For the moment, let "A" represent Ba^{+2}, let "B" represent F^-, let "x" represent the number of moles of Ba^{+2} that appear, and "y" represent the number of moles of fluoride ions. The dissolution of any salt can be explained algebraically as follows:

$$A_xB_y \rightleftharpoons xA^{+y} + yBm^{+x} \qquad [Eq.\ 2\text{--}20]$$

The solubility product equation is:

Chapter 2: The Chemistry of Natural Water Supplies

$$K_{sp} = [A]^x + [B]^y \qquad [Eq.\ 2\text{-}21]$$

So, if we wished to determine the amount of barium fluoride that goes into solution, where $K_{sp} = 1.7 \times 10^{-6}$, we could solve it in a similar manner to the earlier example of barium sulfate. However, in this reaction, the dissolution of BaF_2 produces two moles of fluoride ion, as opposed to just one mole of sulfate from the previous example. So, for every x moles of Ba^{+2} in solution, we have $2x$ moles of fluoride. From Equation 2–19 then, we have the following:

$$K_{sp} = x \cdot 2x^2$$

$$K_{sp} = 4x^3$$

This is a bit more complicated than the previous example, but nonetheless this is how the solubility product is defined. By the way, solving for x gives a value of 7.55×10^{-3} moles per liter. I will leave it to the reader to check my algebra.

Conclusion

This chapter should provide you with a better insight on the chemistry of natural waters, and should also show why water in its natural form cannot be used in steam generating systems. Even trace amounts of impurities can cause enormous difficulties. Since we know that water must be much cleaner before being introduced to a boiler, I have devoted chapter 3 to a general discussion of makeup treatment chemistry. This will put us in position to understand the fouling and corrosion mechanisms outlined in chapter 4. So often these are caused by seemingly minor contamination.

MAKEUP WATER TREATMENT | 3

Introduction

There once was a television commercial in which the first car in a long line of slow-moving traffic was belching smoke. All of the drivers behind were weaving back and forth so that they could see ahead. When a passenger in the offending car asked the driver about his brand of motor oil, the driver growled "Motor oil is motor oil." An announcer came on and said "Motor oil is definitely not motor oil!" I use this example because I have been to plants where operations managers have almost literally said "Water is water." To paraphrase the announcer, "Water is definitely not water!" The natural waters outlined in chapter 2 would tear up most boilers in no time at all. Even at much lower concentrations, impurities can cause great difficulties. We will examine many of these mechanisms in chapter 4, but before doing so let's look at the processes by which water is purified to steam generation quality.

Pretreatment

The production of high-purity water from natural supplies is a step-wise procedure. Most impurities are initially taken out by such tried-and-true processes as settling, clarification, and filtration. Then come the final purification stages, most notably reverse osmosis (RO) and ion exchange (IX). Recently, these methods have been supplemented by newer technologies. These include the membrane techniques of micro- and ultrafiltration, the electrical/membrane process of electrodialysis reversal, and the electrical/ion exchange process known as electrodeionization.

Traditional Pretreatment Concepts

Pretreatment requirements for makeup water are in part dependent on the type of raw water supply. Surface waters, which almost always contain suspended solids, usually must be clarified and/or filtered before further treatment. Groundwaters may be virtually free of suspended solids, but may contain much higher levels of dissolved minerals. Microbiological fouling can be a problem in systems supplied by surface water or groundwater.

Figure 3-1 shows a very common and traditional treatment scheme for a surface water supply. Biocide feed is first, followed by clarification and then filtration. A process that is often integrated with the clarifica-

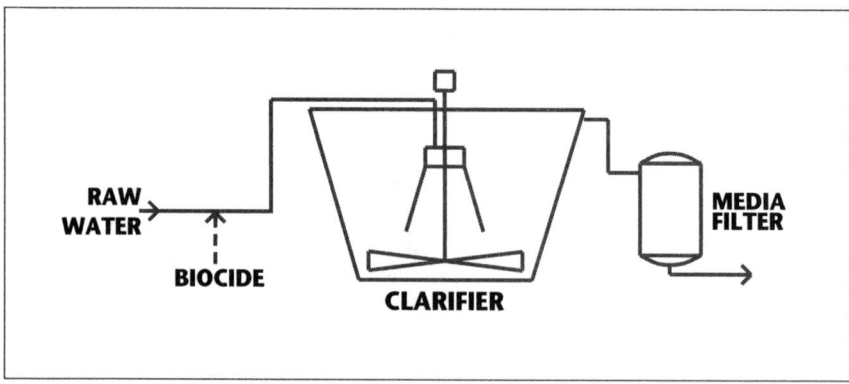

Fig. 3-1   *Outline of a Common Surface Water Pretreatment Scheme.*

tion stage is cold lime softening for hardness reduction. Let's go through the steps shown in Figure 3–1 to learn more about the chemistry of this pretreatment system.

Biocide

Microbiocide programs are covered extensively in chapter 7 (cooling water treatment), and it is there that we will review biocide chemistry in detail. However, makeup water systems can also suffer from microbiological fouling. Although surface waters pick up microbes naturally, groundwaters will also do so if they are exposed to the atmosphere. A common instance is when the water is aerated to precipitate dissolved metals or remove harmful gases. The microbes, if left untreated, will multiply and foul filter media, ion exchange resins, and RO membranes. Fouling restricts water flow and may either directly or indirectly introduce contaminants to the water treatment effluent. One particular type of RO membrane can be attacked and destroyed by microorganisms.

Early microbial treatment of the influent water kills microbes before they have a chance to multiply. The dead organisms come out in the clarifier and filters. The most popular treatment chemical for many years was chlorine gas, which is typically fed into the raw water through an eductor system. Chlorine gas combines with water to produce hypochlorous acid (HOCl), which is the killing agent.

$$Cl_2 + H_2O \leftrightarrows HOCl + HCl \qquad [Eq. 3-1]$$

Chlorine not only reacts with microbes but also with other constituents including ammonia, organics, and, in the case of groundwater supplies, dissolved iron and manganese. These all contribute to the chlorine demand. If an insufficient amount of chlorine is fed to satisfy all demand, microbes may survive and thrive. Thus, it is wise to maintain a free chlorine residual throughout the pretreatment system. A recommended range is 0.1 to 0.5 ppm.

Concerns about the safety of chlorine gas have generated a movement towards less hazardous compounds. A popular alternative is liquid sodium hypochlorite (NaOCl). NaOCl in 5.25% concentration is common household bleach. Industrial solutions typically range from 12% to 15% concentration. Two disadvantages of hypochlorite are that it is several times more expensive than gaseous chlorine, and it may be less effective. When gaseous chlorine is added to water, it tends to lower the pH due to the formation of the two acids shown in Equation 3–1. With sodium hypochlorite, the pH tends to rise. As pH goes up, HOCl dissociates according to the following reaction:

$$HOCl \leftrightharpoons H^+ + OCl^- \qquad [Eq.\ 3\text{–}2]$$

OCl⁻ is a much weaker biocide than HOCl.

Some up-and-coming chemical replacements include solid chlorine donors and bromine. These are discussed in chapter 7. Ultraviolet light is another option, and is most often used in conjunction with other biocides. Learning Aide 3–1 outlines such a combination biocide system at an electronic parts manufacturing plant.

Chlorine and other oxidants will attack certain types of RO membranes and also ion exchange resins. Techniques for removing oxidants are outlined in a later section of this chapter.

Clarification

The suspended solids in surface water supplies would quickly foul RO membranes or ion exchange resin beds. Clarification is a traditional method for removing suspended materials. Figure 3–2 shows the generic design of a solids-contact clarifier. We will follow the clarification process through this unit.

The process takes several stages. First is coagulation. Suspended solids are very small and generally carry a negative electrical charge, which keeps the particles separated. A key step in clarification is charge neutralization. In the diagram shown, this takes place in the central column of the clarifier. This is the rapid mix chamber, where the incoming water and coagulating agent first make contact. The coagulant, which may be inorganic or organic, produces positively charged ions or molecules that neutralize the negative surface charge of the suspended solids. The most popular inorganic coagulants include:

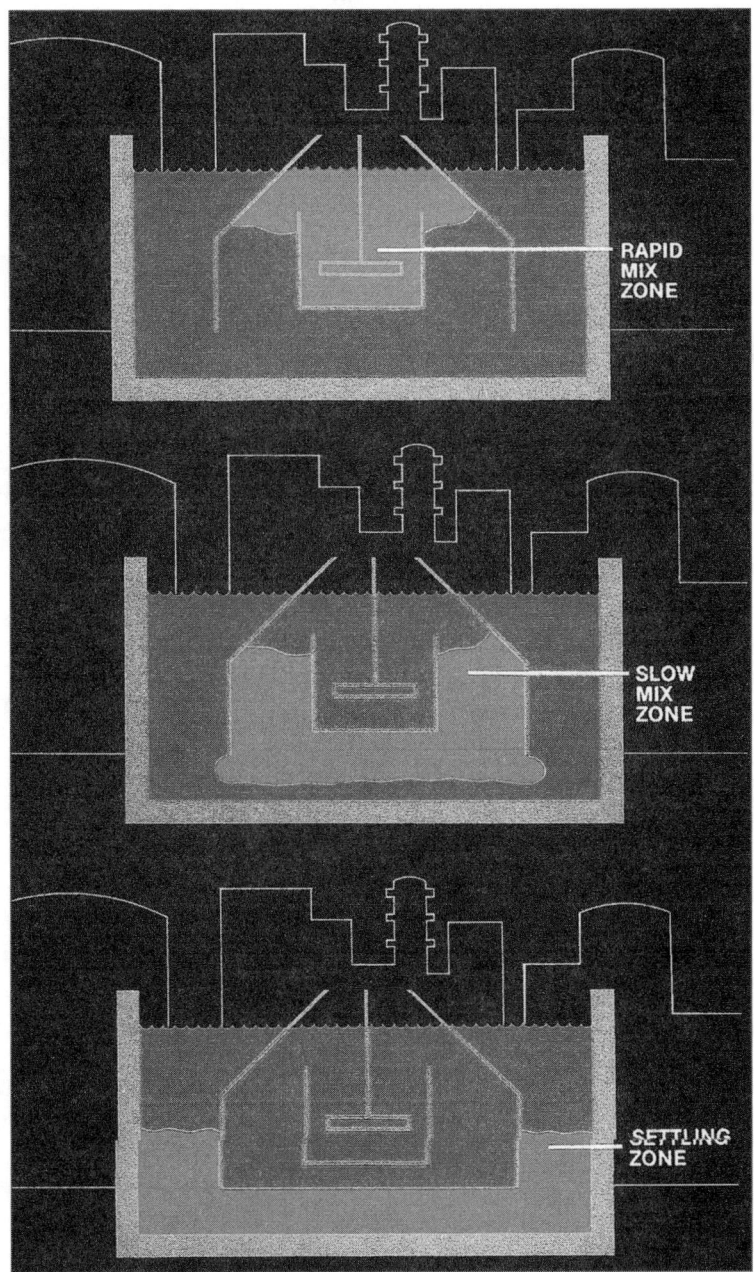

Fig. 3–2 *Outline of Mixing and Settling Zones in a Clarifier.*
Source: Betz Handbook of Industrial Water Conditioning, Ninth Edition. Betz Dearborn is a division of Hercules, Inc.

- Aluminum sulfate, $Al_2(SO_4)_3 \cdot 18H_2O$
- Sodium aluminate, $Na_2Al_2O_4$
- Ferric sulfate, $Fe_2(SO_4) \cdot 9H_2O$
- Ferric chloride, $FeCl_3$

A typical reaction is:

$$Al_2(SO_4)_3 + 6H_2O \rightarrow 2Al(OH)_3\downarrow + 3H_2SO_4 \quad [Eq.\ 3\text{--}3]$$

This Equation illustrates two important aspects about the chemistry. First, the aluminum ions do not exist by themselves but combine with water to form a gelatinous, hydrated precipitate. This aids settling as well as charge neutralization. Second, the reaction of these salts with water influences pH. Alum and the iron salts tend to lower the pH of the solution due to the generation of sulfuric acid. Sodium aluminate has the opposite effect.

$$Na_2Al_2O_4 + 4H_2O \rightarrow 2Al(OH)_3\downarrow + 2NaOH \quad [Eq.\ 3\text{--}4]$$

Aluminum salts function most effectively over a pH range of about 5.5 to 8.0. Iron salts, which also generate hydroxide precipitates, are effective over a broader pH range, typically between 5.0 and 11.0. They are more efficient than aluminum in systems that operate at an alkaline pH of above 8.

The advantage of inorganic salts is that they are common, inexpensive chemicals. Sometimes, though, the salts may not produce the desired effluent quality, and organic coagulants may be needed as a supplement or as the sole coagulating agent. Organic coagulants are typically moderate length polymers with a molecular weight of 50,000 or less. Along the length of the polymer, functional groups are attached to the carbon atoms. A common cationic functional group is the

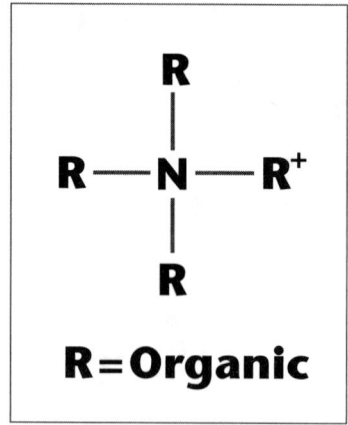

Fig. 3–3 *Amine General Structure*

quaternary amine, whose general structure is outlined in Figure 3–3. The amine group, with its positive charge, is the active agent. Because each coagulant molecule contains many active sites, dosages of a few ppm are generally sufficient for charge neutralization.

It is important that the coagulant be rapidly dispersed into the incoming feed so that charge neutralization occurs quickly. Thus, the rapid mix zone, which by design promotes coagulant-solids contact. As the negative charges on the suspended solids are neutralized, the particles draw closer and combine to form larger solids. At this point an additional flocculating treatment may be needed, as a coagulant alone may not cause the particles to grow large enough to settle rapidly. Whereas coagulants are small molecules, flocculants are long-chain organics, whose molecular weights can reach as high as 50,000,000. These molecules attract coagulated particles. Nonionic or anionic polymers are the most effective flocculants. Polymers with amide functional groups (Figure 3–4) are common nonionic flocculants, while polymers with carboxylate functional groups (Figure 3–5) are the anionic version.

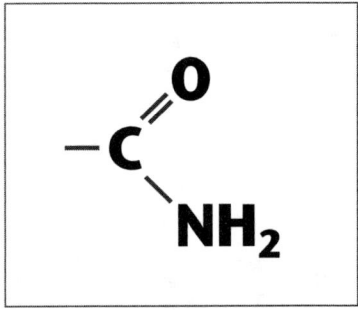

Fig. 3–4 Non-ionic Functional Group

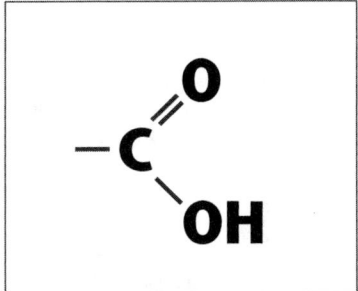

Fig. 3–5 An-ionic Functional Group

Flocs are rather fragile, so this process must be carried out in a relatively calm environment. As Figure 3–2 illustrates, the rapid mix zone is small and the influent water quickly flows from it into the much larger central cone where velocities are lower. This reduces the agitation of the water and allows flocs to develop. In some clarifiers, such as those shown in Figures 3–6 and 3–7, mixers provide gentle agitation to help ensure that the flocculant and suspended particles interact well.

The flow eventually proceeds from the mixing zone to the outer circumference of the clarifier. The fluid flowing up and out of the clarifier

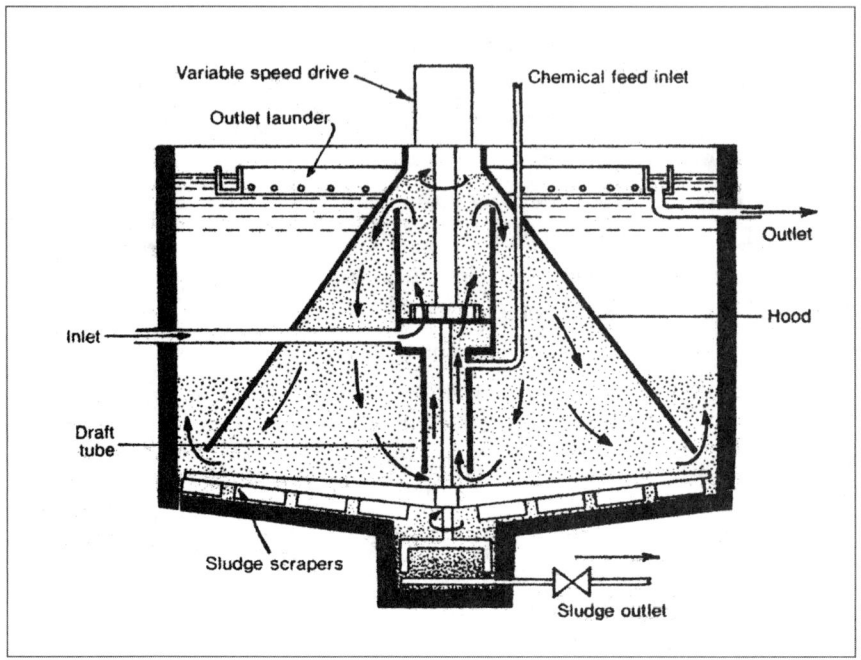

Fig. 3-6 *Clarifier Arrangement I*
Source: Drew Principles of Industrial Water Treatment. Drew Industrial is a subdivision of Ashland Chemical Company.

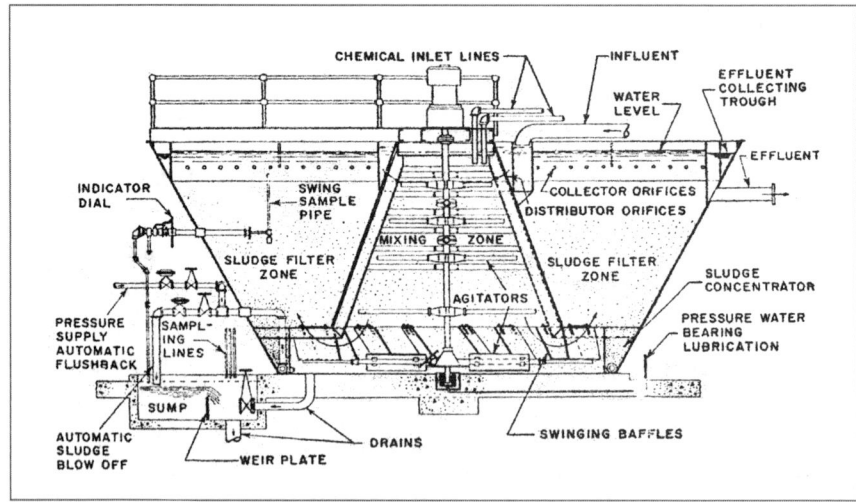

Fig. 3-7 *Clarifier Arrangement II*
Source: Drew Principles of Industrial Water Treatment. Drew Industrial is a subdivision of Ashland Chemical Company.

establishes a blanket or layer of flocculated solids. The blanket traps the newly formed flocs that emerge from the mixing zone. This is where solids-contact clarifiers get their name. Maintenance of a good sludge blanket is important, as Practical Example 3–1 illustrates.

The rate at which water flows through the main body of the clarifier is known as the rise rate. A typical range of rise rate for the solids-contact clarifiers shown in earlier illustrations is 0.5 to 1.5 gpm per square foot of surface area. Retention times may range from 1 to 4 hours.

Clarification produces two streams; the clear water that flows from the surface of the unit and sludge that must be periodically withdrawn from the bottom of the vessel. Where flow and suspended solids concentrations are fairly consistent, sludge bleed off may be set by an automatic timer. The frequency may only need to be changed for seasonal or weather-related effects. In other cases, the operator may have to pay close attention to factors such as blanket level and sludge density, and adjust the bleed off more often.

Figure 3–8 shows a another clarifier, in which flow is more horizontal in nature. The rapid mix, flocculation, and settling zones are clearly

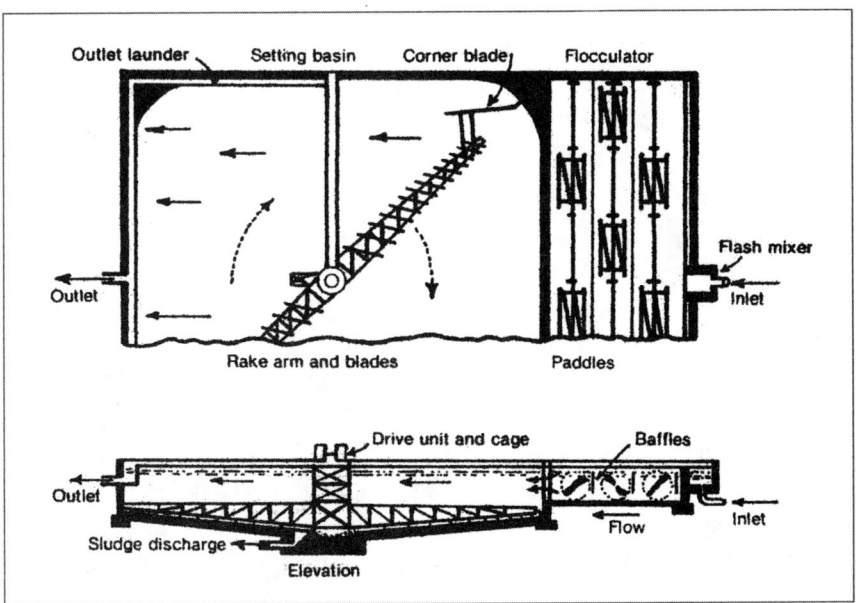

Fig. 3–8 *Clarifier Arrangement III*
Source: Drew Principles of Industrial Water Treatment. Drew Industrial is a subdivision of Ashland Chemical Company.

defined. While in-ground circular clarifiers are often specified for large applications, clarification/softening of water at flow rates of up to 1000 gpm or even more can be accomplished in individual or parallel package clarifiers. These units are of rectangular shape, but are smaller and of different design than that shown in Figure 3–8. One of the primary differences is that the main body of the clarifier is equipped with a series of inclined plates or tubes. The water flows upwards along the partitions to the outlet. The interaction of particles flowing upwards with heavier particles settling downwards enhances the contact between newly-formed floc and the heavier solids. Residence times in these clarifiers, often referred to as Lamella type, may be much shorter than in conventional solids-contact clarifiers. Package clarifiers are usually less expensive than in-ground units, where installation costs may be significant. Systems of less than 600 gpm or so may be shipped as one unit. Even large systems can be shipped in just a few pieces and be field erected.

Many factors may upset the performance of a clarifier. A quite obvious factor is change in flow. A sudden increase in flow will increase the rise rate and cause the blanket to expand. An increase in solids loading is another factor. This increases the load on the clarifier, requiring greater chemical dosages. A decrease in solids concentration has the opposite effect. A change in water chemistry for whatever reason can also create problems. Chemistry fluctuations may occur due to natural causes or manmade changes. Practical Examples 3–2 and 3–3 illustrate an example of each.

The analytical standard by which clarifier performance is often monitored is turbidity. When a beam of light is shown through water, suspended solids diffuse or deflect the rays. This can be measured, and the answer given in units of turbidity. The most common measurement is nephelometric turbidity units or NTU, and instruments, both benchtop and on-line, are available to determine the NTU of many waters. "Clean" surface waters can have a turbidity of less than 20 NTU. Average waters range between about 20 and 60 NTU, while those with a lot of suspended solids will be over 60 NTU. A properly operating clarifier should reduce turbidity to below 10 NTU and perhaps as low as about 2 NTU. Surprisingly, the clarifier may not perform efficiently if the influent turbidity is very low. It may be necessary to add a bulking agent to improve performance. In these cases, if the raw water suspended solids concentra-

CHAPTER 3: MAKEUP WATER TREATMENT

tion is fairly consistent, clarification may not be needed at all. Filtration, with upstream coagulant feed, may be satisfactory.

It is usually impossible to determine from water analyses alone the correct dosages and type of chemicals that will be most effective. A good initial determination is possible through jar testing, in which a water treatment firm will come in, collect samples of the actual process stream and then test various coagulants and flocculants in one liter samples of the water that are agitated with mechanical stirrers. Jar testing can determine effective chemicals and approximate dosages. This data can then be used as a starting point for full-scale treatment. After that, observation, chemical analyses, and common sense will guide the operator.

Practical Example 3-1

The industrial wastewater treatment system at my former utility operates very similarly to traditional makeup water clarification. Power plant waste streams such as boiler blowdown, soot blower drips, and the discharge from floor drains and roof drains all flow to the plant. The water first passes through a settling basin to allow large particles to settle, is treated with a cationic coagulant in a rapid mix chamber, and then flows to three, parallel, cone-shaped solids-contact clarifiers. The clarifier effluent discharges directly to a recreational lake, while the sludge discharges to the plant ash ponds.

I came in one Monday morning to find the wastewater plant operator, a normally calm person, beside himself with frustration. A relief operator over the weekend had turned on the sludge removal system and then forgot to turn it off. When the regular operator came in Monday morning, the sludge blankets in each clarifier were gone, kaput, vanished. Turbidity of the effluent was quite high, as the newly-formed flocs had no material on which to collect. The clarifiers did not fully recover for several days.

Practical Example 3-2

Another example from my former plant, which is a combined electric and water utility. Makeup water to the power plant is supplied from the water treatment facility. Raw water is taken from the lake. Treatment includes clarification. One of the treatment chemicals is activated carbon, which adsorbs organic pesticides that wash into the lake from farm fields. Each spring, the operators must increase the dosage rate of activated carbon due to water runoff. Ammonia levels in the water also increase during this time period due to fertilizer runoff. Heavy rains may stir up many solids including collodial silica. The effects of this troublesome contaminant are outlined in Practical Example 3–4 later on.

Practical Example 3-3

This example is taken from a manufacturing facility, and although it involves the non-sanitary wastewater treatment system, the example still illustrates factors that affect clarifier performance. The plant's two clarifiers are similar to the design shown in Figure 3–2. Coagulation occurs naturally through actions of microbes. For years, the clarifiers handled water whose initial low pH had been neutralized with calcium hydroxide $Ca(OH)_2$ followed by sodium hydroxide. The top of the sludge blanket in each clarifier typically steadied out around 8 feet below the water surface.

A change in chemical treatment replaced calcium hydroxide with magnesium hydroxide. Immediately the blankets in both clarifiers began to rise until sometimes the tops of the blankets were just below the water surface. Suspended solids developed a fuzzy look resembling pin floc. Although the precise mechanism for this problem has not yet been determined, the literature indicates that the magnesium from the pre-neutralization step is being reconverted to magnesium hydroxide during the caustic neutralization stage. Very small particles of magnesium hydroxide are thought to adsorb or entrap air and float, thus expanding the sludge beds. Regardless of the exact mechanism, this example illustrates how seemingly innocuous changes can affect pretreatment operation.

Filtration

Although clarification removes most suspended solids, enough still carry over from the clarifier to foul RO membranes or ion exchange beds. Filtration is the next step in the pretreatment process. The most common filters for industrial makeup treatment are media-type, pressure filters, in which the feed water passes through layers of graded material that trap particles. Several types of media are practical, and often they work best in combination. The simplest media is sand, and in some filters this may be the only material other than a support bed of gravel. Even though filtration appears to be a simple process, certain aspects must be taken into account when designing or operating filters. Most solids are removed in the first few inches of the bed. If only a single grade of media is used, the top portion of the bed may become blinded with material. Better is two or more grades or types of media, with the coarsest, less dense material on top followed by finer and more dense grades in succession. The coarse media removes the largest solids, while the finer media remove smaller particles and polish the water. Several materials besides or in combination with sand serve as suitable filter media. These include anthracite and garnet. Table 3–1 illustrates the typical size ranges for these media, and Learning Aide 3–2 outlines the media specifications and bed depths for an actual set of filters at an electric utility.

Pressure filters are usually designed to operate within a range of 2 to 4 gpm per square foot of filter media. Thus, a filter with a diameter of 10

Filter Media Sizes

Media	Effective Size, mm (in.)	Specific Gravity
Anthracite	0.7–1.7 (0.03–0.07)	1.4
Sand	0.3–0.7 (0.01–0.03)	2.6
Garnet	0.4–0.6 (0.016–0.024)	3.8
Magnetite	0.3–0.5 (0.01–0.02)	4.9

Table 3–1 *Typical Size Ranges for Filter Media*

feet could potentially treat a flow ranging from 157 to 314 gpm. Backwash rates are typically two or more times greater than that of the operating flow rate. This allows for good expansion of the filter bed during the washing process. The backwash rate must not be set too high, as this could blow some of the lighter media out of the filter. Changes in water temperature influence the water density, and modifications to backwash flow rates are often required at different times of the year. In winter, the flow rate may need to be lowered, with the opposite modification in summer. Backwashing is an extremely important step. While excessive backwashing can cause loss of media, insufficient backwashing can allow the accumulated solids to filter down into the media and form mudballs. The mudballs then affect subsequent operation and backwash cycles of the filter.

A properly operating filter will remove particles down to about 25 microns in size, and should produce water with a turbidity of less than 1 NTU, and sometimes down to 0.1 NTU.

Where surface supplies have few suspended solids, filtration alone may be suitable for preparing the raw water for further treatment. In many cases, a cationic coagulant is fed just ahead of the filters to initiate particle growth. Coagulant feed too far upstream of the filters could cause blinding. This type of treatment should be viewed very cautiously if the downstream equipment includes a reverse osmosis unit. The most popular type of RO membrane, the thin-film-composite, has a negative surface charge. Any cationic polymer that passes through the filter bed will foul the RO membranes.

Water Softening in the Pretreatment Process

While clarification and filtration remove suspended solids, by themselves they have virtually no effect on dissolved solids including the hardness ions. Heavy loadings of dissolved ions can cause short demineralizer run lengths, scaling and fouling of RO membranes, and other problems. A common method to reduce hardness concentrations is through lime softening in the clarifier.

In chapter 2, we examined the typical dissolved ions in raw water supplies. The most predominant cations include sodium, potassium, calcium and magnesium. The predominant anions include bicarbonate, chloride, and sulfate. Since waters are electrically neutral, cations and anions stay within close proximity to maintain neutrality. If a sample of the water were to be evaporated, one would find ions pairing off in certain orders of preference. Most importantly, calcium and magnesium would first combine with bicarbonate ions, followed by sulfate, then chloride. This ion pairing has a large influence on the softening process. Consider a water supply in which more calcium and magnesium ions exist than equivalent bicarbonate ions. The calcium and magnesium that associate with bicarbonate are known as carbonate hardness, while the excess that associate with sulfate and chloride ions are known as non-carbonate hardness.

Removal of hardness by softening relies upon the difference in solubilities of various calcium and magnesium compounds. This is clearly illustrated by the primary softening reactions that remove calcium and magnesium carbonate hardness from water.

$$Ca(HCO_3)_2 + Ca(OH)_2 \rightarrow 2CaCO_3\downarrow + 2H_2O \quad [Eq.\ 3\text{--}5]$$

$$Mg(HCO_3)_2 + 2Ca(OH)_2 \rightarrow Mg(OH)_2\downarrow + 2CaCO_3\downarrow + 2H_2O \quad [Eq.\ 3\text{--}6]$$

The addition of lime raises the pH and converts the bicarbonate to carbonate ions. As we learned earlier, $CaCO_3$ is one of the least soluble of the common hardness salts. With proper operation, a clarifier can reduce calcium to 35 to 40 ppm (as $CaCO_3$). Magnesium reduction occurs by a different process. As pH rises to 10 and above, magnesium precipitates directly with hydroxide ions. Note that twice as much lime is needed to precipitate magnesium. Figures 3–9 and 3–10 show the relationship in solubility to that of pH for calcium and magnesium.

Lime treatment will also reduce magnesium non-carbonate hardness due to the increase in pH and formation of magnesium hydroxide.

$$Mg(SO_4)_2 + Ca(OH)_2 \rightarrow Mg(OH)_2\downarrow + CaSO_4 \quad [Eq.\ 3\text{--}7]$$

Lime treatment alone will not reduce calcium non-carbonate hardness. Supplemental carbonate alkalinity is needed to react with calcium.

$$CaSO_4 + Na_2CO_3 \rightarrow CaCO_3\downarrow + Na_2SO_4 \quad [Eq.\ 3\text{--}8]$$

Some silica removal may also take place during the softening process if magnesium is present. The silica precipitates as magnesium silicate. If the magnesium concentration is low, supplemental magnesium salts may be added to enhance silica removal. Figure 3–11 illustrates the silica removal that is possible in lime softened systems.

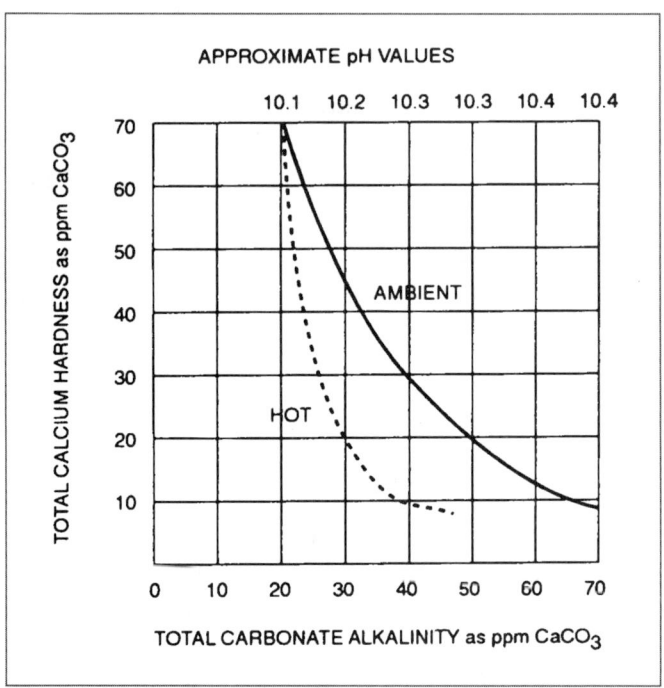

Fig. 3–9   *Calcium Removal vs. Carbonate Alkalinity*
Source: Betz Handbook of Industrial Water Conditioning, Ninth Edition. Betz Dearborn is a division of Hercules, Inc.

A number of literature sources provide calculations of lime and carbonate feed required for waters of varying constituency. One good reference is U.S. Filter's Water And Waste Treatment Data Book.

For hardness removal from small volume flows, hot-lime softening is a viable treatment. This method has been fairly popular for industrial applications. Hot-lime softeners employ the same chemistry as cold-lime units, with the exception that the incoming water is heated with steam or some other source in a closed vessel. This enhances the chemical reactions, in part by lowering the solubility of the precipitates. Hot-lime softening can reduce hardness and silica to the following levels:

- Calcium — 15 ppm
- Magnesium — 5 ppm
- Silica — 1–2 ppm

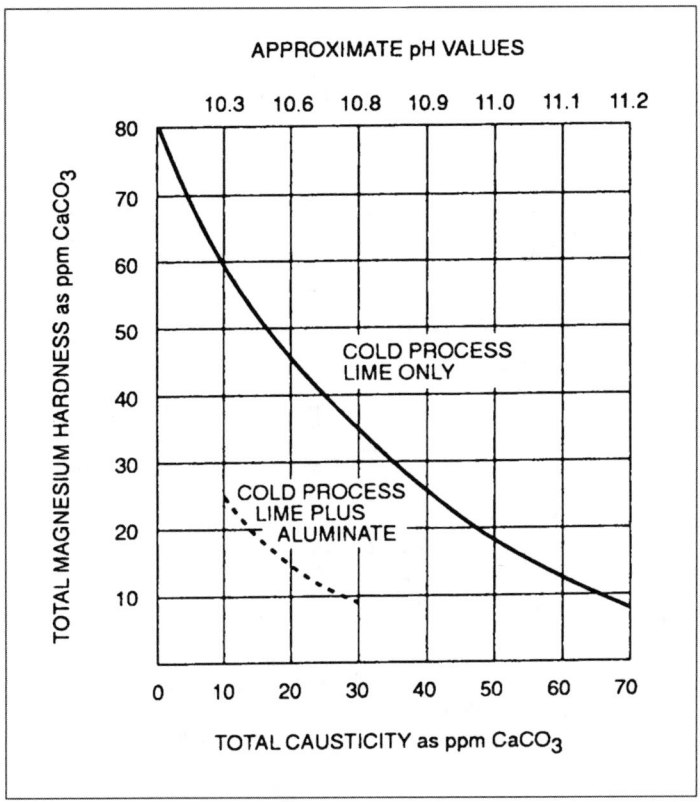

Fig. 3–10 *Magnesium Removal vs. pH Source*
Source: Betz Handbook of Industrial Water Conditioning, Ninth Edition. Betz Dearborn is a division of Hercules, Inc.

Post-Treatment of Pre-Treatment

This may seem like a funny title, but its purpose should become clear. The effluent from our example system above may not yet be ready for polishing, especially if it still contains residual oxidizing biocides. Oxidizers will attack and degrade ion exchange resins and especially thin-film-composite RO membranes. Two methods are common for removal of oxidizers. The first is a fixed treatment process using activated carbon and the second involves chemical treatment.

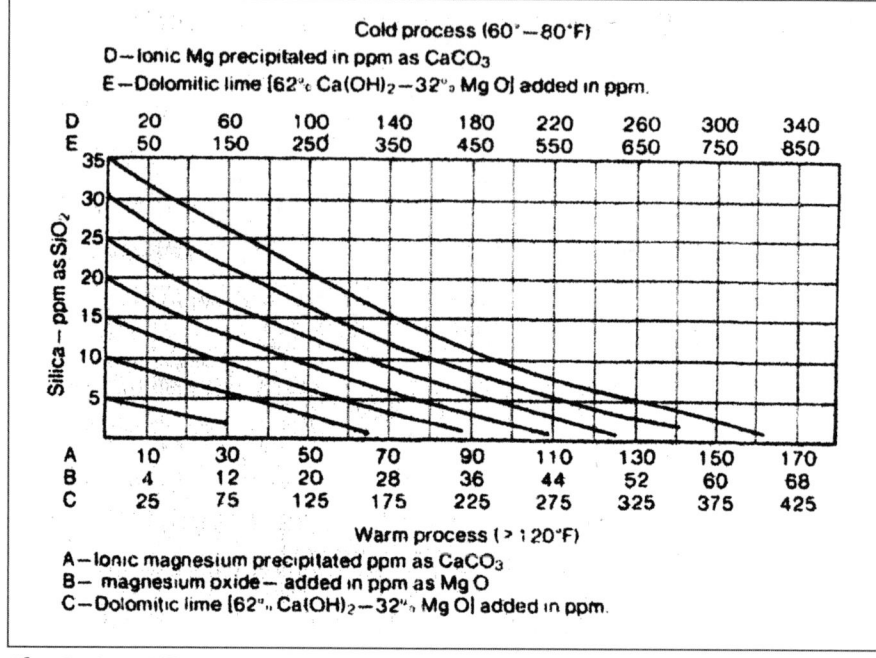

Fig. 3–11 *Silica Reduction Possible in the Softening Process*
Source: Betz Handbook of Industrial Water Conditioning, Ninth Edition. Betz Dearborn is a division of Hercules, Inc.

Activated Carbon Filtration

Activated carbon is produced by roasting organic materials in an oxygen-free atmosphere. The most common starting materials include coal and coconut shells. The process generates a finely-grained product with tremendous adsorptive capacity. Carbon filters somewhat resemble multimedia filters in design and operation, except that they do not serve as a primary filtering device. They will, however, remove some of the large, natural organics found in surface supplies. Activated carbon removes chlorine and other oxidizers due to direct reaction.

$$C + HOCl \rightarrow CO + HCl \qquad [Eq. \; 3-9]$$

These reactions take place within the top layers of the bed, which allows the remaining portion of the bed to become an excellent breeding ground for bacteria that survive the biocide treatment. This can cause severe problems in downstream equipment. One method to combat carbon bed microbiological fouling is to equip the filter with an auxiliary steam line taken from a plant source. Periodic introduction of steam to the filter will sterilize the activated carbon. If sterilization of the bed is not possible, the carbon should be replaced on a fairly frequent interval. Six months is a reasonable guideline.

An alternative to the fixed method of activated carbon filtration is chemical reduction of residual biocides. The most common reductants include sodium sulfite (Na_2SO_3), sodium bisulfite ($NaHSO_3$), and even sulfur dioxide (SO_2). The following equation typifies this chemistry:

$$Na_2SO_3 + HOCl \rightarrow Na_2SO_4 + HCl \qquad [Eq.\ 3\text{--}10]$$

Sodium bisulfite and sodium sulfite may be purchased in liquid form and fed through a simple metering pump. Sulfur dioxide is a gas.

Even this type of treatment may still allow microbiological fouling downstream of the reductant injection point. We will examine a case history of microbiological fouling in the section on reverse osmosis. We will also look at other pretreatment requirements for RO units, as clarification/filtration may not be totally satisfactory.

Extra Filtration for RO Units

Even a well-run multimedia filter does not normally reduce particulate load enough to protect RO membranes from fouling. Pretreatment for reverse osmosis systems usually includes 10-micron and or 5-micron depth filtration directly ahead of the unit. The RO vendor will supply this subsystem as part of the RO skid.

Alternatives to Clarification and Conventional Filtration

Recent advances in technology offer alternatives to traditional suspended solids removal techniques. Two techniques that are gaining acceptance are microfiltration (MF) and ultrafiltration (UF). These are cross-flow filtration processes, in which suspended solids are removed from water as it passes along and through a synthetic membrane filled with microscopic channels. We will learn more about these membranes in the section on reverse osmosis.

Pretreatment of Groundwaters

Groundwaters as a rule contain very few suspended solids, which often eliminates the need for clarification. However, as we learned in chapter 2, dissolved solids concentrations may be quite a bit higher. These can include all of the major ions including hardness, bicarbonate, chloride, sulfate, silica, iron, and manganese. The water may also contain dissolved gases including carbon dioxide and, on occasion, hydrogen sulfide. While it may not be practical to remove many of the dissolved ions before the final purification steps, it is possible to remove iron, manganese, and dissolved gases in the pretreatment system. Figure 3–12 shows the groundwater pretreatment schematic at a manufacturing facility. Raw water is

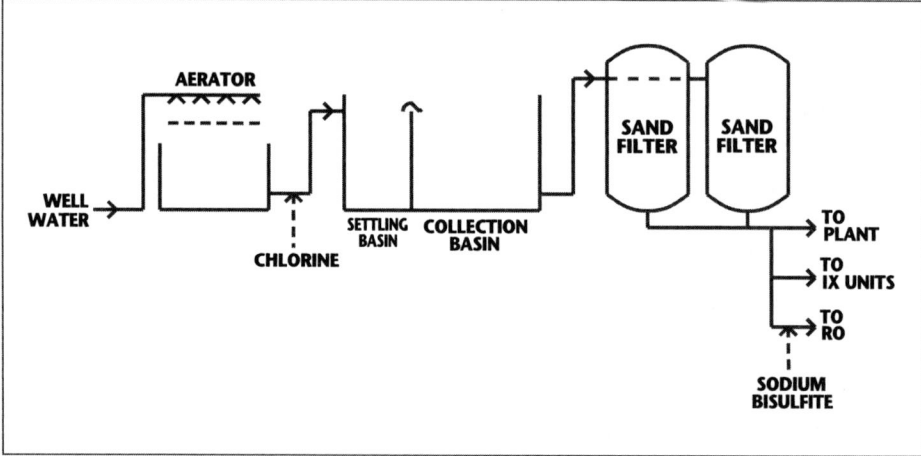

Fig. 3–12 *Example of a Groundwater Pretreatment System*

first passed through spray/tray aerators to begin the oxidation of iron and manganese, and to liberate dissolved gases. Next comes chlorine feed both to prevent microbiological growth and to assist the oxidation of manganese. The stream passes through a settling chamber where much of the iron and manganese fall out. Indeed, after several months of operation one can see a mound of the oxides in the settling basin. The water is then pumped through multi-grade sand filters to remove residual iron and manganese oxides. The sand in these filters will turn dark over time as it removes the oxides. The final step is dechlorination with sodium bisulfite, upon which the water is processed in an RO unit and a parallel ion exchange system for feed to four industrial boilers. There is one note of interest about the groundwater supply. It comes from three wells all located within a few hundred feet of each other near a major river. All wells are of similar depth. Water quality from the first two wells is nearly identical but water from the third well shows consistently higher dissolved solids. Seasonal changes in the river quantity and quality have an effect on all three wells.

One other example of interest. The original aerator was equipped with coke-filled trays. Coke acts as a catalyst in the oxidation of iron and manganese. However, the coke plugged so rapidly with oxides that frequent replacement was necessary. Plant personnel removed the coke as an experiment. Subsequent analyses indicated that free-fall aeration plus chlorination were still effective in oxidizing most of the iron and manganese.

Final Polishing Techniques

Clarification/softening or alternative techniques remove most of the suspended solids and many of the dissolved solids from the incoming raw water. However, the water quality is still unsuitable for most boilers, and certainly for any that produce steam for electrical power. Polishing is a must.

High-Purity Makeup Treatment Methods

The most important and common makeup treatment methods are ion exchange and reverse osmosis. These technologies are mature. Other techniques that have begun to emerge and show promise include electrodialysis reversal (EDR) and electrodeionization (EDI). The membrane technologies of microfiltration (MF) and ultrafiltration (UF) are becoming popular for pretreatment, as we shall see.

Ion Exchange

Ion exchange actually occurs in nature. Water passing through soil will exchange ions, principally cations. The roots of the industrial ion exchange process can be traced back to the early 1900s when the German scientist, Gans, used synthetic zeolites (sodium aluminum silicates) to exchange calcium and magnesium ions for sodium. This discovery led to the emergence of water softeners. In the 1930s and 1940s, researchers developed synthetic, polymer-based ion exchange resins, and these revolutionized the industry.

The efficiency of the ion exchange process starts with the exchange media itself. Important qualities of ion exchange media include:

- It must have the maximum number of exchange sites.
- It must have structural integrity.
- Easy loading of the media into the reaction vessel is required.
- It must be of reasonable cost.
- It must be capable of being regenerated efficiently.

The design that has proven most practical for ion exchange resins is that of small spherical plastic beads. The beads behave somewhat like a fluid, which makes them easier to handle and load/unload into exchange vessels. By far, the most common material from which ion exchange resins are fabricated is polystyrene cross-linked with divinylbenzene (Figure 3–13), although acrylics are used in some instances. The beads are manufactured in an emulsion process, which allows the polymer to form spheres. The fabrication process and degree of cross-linking can be controlled to vary the porosity within, and structural strength of, the beads. Two kinds of resins are most common. These are defined as gel-type and macroporous (macroreticular). The former have a relatively low divinylbenzene cross-linkage of perhaps 8 percent or so. Gel-type beads do not have discrete pores, although in water they will swell and allow the passage of ions through the structure. Macroreticular resins have cross-linkages of 12 to 20 percent, which impart a greater rigidity and strength to the resin, and give the resin a porous structure. Resins of different porosity and strength are useful in different environments. For example,

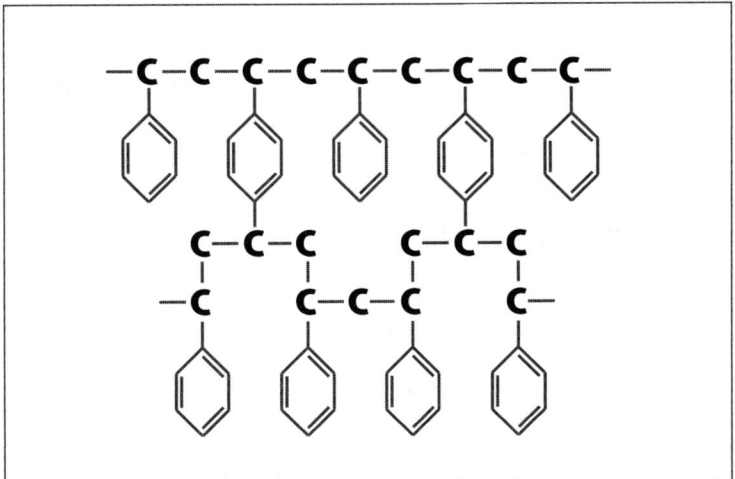

Fig. 3-13 *Organic backbone of Common ion Exchange Resins*

macroporous resins are more durable in heavy-duty applications. The gel-type resins typically contain more exchange sites. Both types are heavily used in the water treatment industry.

Macroreticular resins with large pores and no exchange sites have been developed for removal of organic complexes from raw water. The resin functions by adsorption of the organics on and within the bead. The resin is non-regenerable and is discarded when it reaches exhaustion. These resins show promise for removing relatively small-chain organics from condensate, which are otherwise difficult to treat.

Exchange Groups

When beads are synthesized, no exchange groups are present. These are added afterwards. The principal exchange groups are:

- Sulfonic acid ($SO_3^- H^+$) — Strong acid cation resin
- Carboxylic acid ($COO^- H^+$) — Weak acid cation resin
- Quaternary amine ($CH_2N(CH_3)_3^+ OH^+$) — Strong base anion resin
- Primary, seconday, and tertiary amines (CH_3NH_2) — Weak base anion resin

The exchange groups give the resins their character and define how they will perform when in service. The most popular ion exchange systems for high-purity water treatment utilize strong acid cation and strong base anion resins, although weak acid cation and weak base anion resins can be effective. We will examine the general chemistry of each.

Strong Acid Cation Resin

All of the exchange sites listed above can be likened in performance to their acid or base counterparts. This is most clearly illustrated with a strong acid cation (SAC) resin. SAC resins have an affinity for cations over hydrogen, and will exchange hydrogen for them as water flows through the vessel.

$$\left. \begin{array}{c} Ca \\ Mg \\ Na \end{array} \right| \begin{array}{c} SO_4 \\ Cl \\ HCO_3 \end{array} + \sim C\text{-}SO_3\text{-}H^+ \leftrightarrows \sim C\text{-}SO_3 \begin{array}{c} Ca^{+2} + H_2SO_4 \\ Mg^{+2} + HCl \\ Na^+ + H_2CO_3 \end{array} \quad [Eq.\ 3\text{--}11]$$

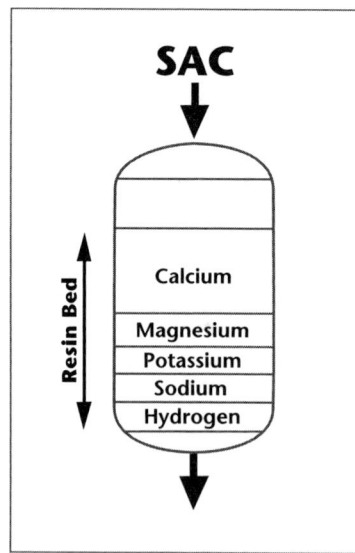

Fig. 3–14 *Selectivity of a Strong Acid Cation Resin*
Courtesy of the Purolite Co., Bala Cynwyd, PA.

As the equations illustrate, cations attach to the exchange sites releasing hydrogen ions. This produces an effluent with a decidedly acidic character, and the pH of a cation vessel effluent may be 2 or lower. SAC resins are known as salt splitters, because the resin actually separates cations from their conjugate anions. Typically, higher charged cations have a greater attraction to the exchange resin. The order of affinity for SAC resin is $Ca^{+2} > Mg^{+2} > Na^+$. As Figure 3–14 shows, during normal operation of an SAC exchanger, the ions tend to form relatively distinct bands within the resin bed. This has a marked influence on the regeneration process, as will be illustrated. Notice that Equation 3–11 uses an equilibrium symbol. This indicates that the resin is not completely efficient. As water flows through the resin, cations will attach, detach, and then reattach. Cations with a greater affinity will dislodge those with lower affinities, which must then find another site. Eventually, enough exchange sites are consumed and ions begin to break through. Sodium has the lowest affinity for the resin and will show up in the effluent first. When, or ideally just before, breakthrough occurs, the resin must be regenerated with acid. We will cover regeneration in greater depth shortly.

SAC resins are also commonly used in water softeners, but in this case the exchange sites contain sodium rather than hydrogen ions. The regenerant is a salt solution.

The performance of a cation bed is influenced by the water flow rate through the bed. Too great a flow rate can potentially cause chanelling or overwhelm the kinetic ability of the resin to exchange ions. Table 3–2 outlines the operating characteristics for a very common cation resin. This data reveals a number of important features and provides practical information on how an exchanger operates. The data shows:

- The total capacity of the resin in unused form is 39.2 kilograins (1000 grains) per cubic foot of material. This gives an idea of its exchange capacity. A grain per gallon is approximately equivalent to 17 ppm of ions as $CaCO_3$. As we shall

Strong Acid Cation Exchanger Specifications

Polymer Structure — Polystyrene cross-linked with dibynlbenzene
Total Capacity Hydrogen Form — 39.2 kgr/ft^3 Swelling (NaRH) — 5%
Standard Operating Conditions

Operation	Rate	Solution	Minutes	Amount
Service	1–5 GPM/ft^3	Influent water		
Backwash	3–5 GPM/ft^3	Influent water	5–20	10–25 gals/ft^3
Regeneration	0.2–0.8 GPM/ft^3	0.5–5% H_2SO_4	30	4–10 lbs/ft^3
Slow rinse	0.2–0.8 GPM/ft^3	Decationized	60	20 gals/ft^3
Fast rinse	1–5 GPM/ft^3	Decationized	60	30 gals/ft^3
Backwash expansion	50–75%			

* Data courtesy of The Purolite Company, Bala Cynwyd, Pennsylvania

Table 3–2 *Strong Acid Cation Exchanger Specifications*

see, once the resin is placed in operation, the operating capacity is reduced by a third or more due to chemistry factors.

- The resin swells in size by about 5% as it is regenerated.
- The optimum service flow rate is 1 to 5 gpm per cubic foot of material. Higher flow rates in particular can lead to reduced performance.
- Backwash flow rates are comparable to service rates. The backwash expansion volume is recommended at 50% to 75%. This is why conventional demineralizers are designed with a lot of freeboard (space between the resin surface and top of the vessel).
- Regeneration rates are several times lower than the service flow rate, which gives the regenerant good contact time.
- Once a regeneration is complete, the unit is slow rinsed for a period of time. The flow rate is the same as the regenerant rate. This gives the residual regenerant a chance to remove additional cations before it is flushed from the unit.
- A fast rinse follows the slow rinse to purge the vessel of acid.
- The recommended regenerant dosage ranges from 4 to 10 lbs per cubic foot. This value will vary depending upon the makeup water quality and the desired effluent. More regenerant typically produces a better regeneration, but it is more costly. A common range for many systems is 5 to 6 lbs/ft^3.

Additional design data is shown in Table 3–3. Let's use this data to see how we might roughly calculate the resin needed for a particular application. Assume that we have to treat 50 gpm of clarifier/filter effluent that has 100 ppm of cations as $CaCO_3$. Let's also say that our calculations show that the resin is only 60% regenerable. Thus, every cubic foot of resin would be able to remove 23.5 kilograins of material. Theoretically, one cubic foot of material could treat 50 gpm of flow for 80 minutes. Obviously, one cubic foot is a ridiculously small amount of resin, so let's refine the calculations assuming a minimum bed depth of 30 inches and a circular cross-sectional flow area of 4 gpm/ft^2. Resin volume equals 31.25 ft^3. Exchange capacity equals 31.25 ft^3 • 23.5 kgr/ft^3 or 734 kilo-

CHAPTER 3: MAKEUP WATER TREATMENT

Guidelines for Good Hydraulic Performance of Ion Exchangers

Parameter	Range of Value
Volumetric flow rate (gpm/ft^3)	0.25 to 5
Cross-sectional flow rate (gpm/ft^3)	4 to 12
Pressure drop per foot of resin (psi/ft.)*	0.1 to 1.5
Minimum bed depth (inches)**	30

* Pressure drop is determined by cross-sectional flow rate and water temperature. Values shown are for 4–10 gpm per sq. ft. flow rates and a temperature range of 40°F to 100°F.

** The book Practical Principles of Ion Exchange Water Treatment (see Bibliography) recommends 9 feet as a maximum practical bed depth.

Table 3-3 *Ion Exchanger Hydraulic Performance Guidelines*

grains. The cation unit could treat our influent for 2500 minutes, or over a day and a half.

Weak Acid Cation Resins

The order of affinity for ions in a weak acid cation (WAC) resin is $H^+ > Ca^{+2} > Mg^{+2} > Na^+$. Thus, WAC exchange sites tend only to release hydrogen ions to an anion, such as bicarbonate, that has a strong affinity for H^+. WAC resins are not true salt splitters. One may ask then, "If a WAC resin does not remove all cations, what is its practicality." The answer is regeneration efficiency. Because a WAC resin has a strong affinity for hydrogen, the resin is much more easily regenerated than a SAC resin. WAC resins can be very effective for treating high-alkalinity waters. Final cation polishing may then be accomplished with SAC resin.

Strong Base Anion Resins

Strong base anion (SBA) resins are the counterpart to SAC resins, and will remove virtually all anions. The resin exchanges hydroxide ions for the anions in the preferential order $SO_4^{-2} > Cl^- > HCO_3^- > HSiO_3^-$ (Figure 3-15). The practicality of ion exchange is illustrated by the reactions that occur when the cation exchanger effluent flows through a SBA exchanger.

FUNDAMENTALS OF STEAM GENERATION CHEMISTRY

$$\left.\begin{array}{l} HCl \\ H_2SO_4 \\ H_2CO_3 \\ H_2SiO_3 \end{array}\right| + CH_2N(CH_3)_3 + OH^- \rightleftharpoons C\text{-}NH_4 + \begin{array}{l} Cl^- \\ SO_4^{-2} \\ CO_3^{-2} \\ SiO_3^- \end{array} + H_2O \quad [Eq.\ 3\text{-}12]$$

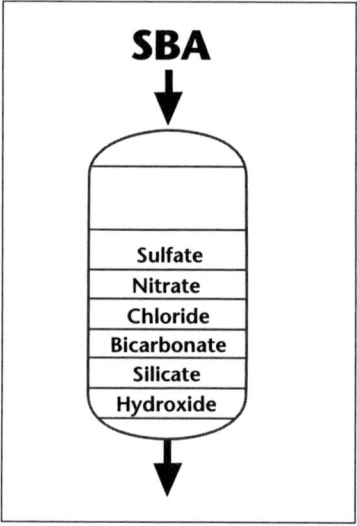

Fig. 3–15 *Selectivity of a Strong Base Anion Resin*
Courtesy of the Purolite Co., Bala Cynwyd, PA.

The final product is water.

Two types of SBA resins are most common, Type I and Type II. Type I resins contain quaternary amine exchange sites as shown in Equation 3–12. Type II contains the quaternary ammonium functional group $(CH_2N(CH_3)_2CH_2CH_2OH)$. Type I resins are more stable at higher temperatures. Type II resins are slightly less basic and can be regenerated a bit more efficiently. Table 3–4 outlines common design characteristics for SBA resins. The reasoning is very similar to that outlined for SAC resins above, so I will not go into it further.

Strong Base Anion Exchanger Specifications

Polymer Structure — Polystyrene cross-linked with dibynlbenzene
Total Capacity Chloride Form — 28.3 kgr/ft³   Swelling (Cl→OH) — 20%
Standard Operating Conditions — Countercurrent Regeneration

Operation	Rate	Solution	Minutes	Amount
Service	1–5 GPM/ft³	Decationized water		
Backwash	2–3 GPM/ft³	Decationized water	10–25	10–25 gals/ft³
Regeneration	0.25–0.5 GPM/ft³	0.5–5% NaOH	30–60	3–10 lbs/ft³
Slow rinse	0.25–0.5 GPM/ft³	Decationized water	30 (approx.)	15–30 gals/ft³
Fast rinse	1–5 GPM/ft³	Decationized water	20 (approx.)	25–45 gals/ft³
Backwash expansion	50–75%			

* Data courtesy of The Purolite Company, Bala Cynwyd, Pennsylvania

Table 3–4 *Strong Base Anion Exchanger Specifications*

Weak Base Anion Resins

Weak base anion (WBA) resins serve a similar function to WAC resins; they remove a portion of the ions so that a downstream SBA resin does not have as heavy a load. The functional group of the WBA resin is a weak base (primary, secondary or tertiary amine), the resin removes weak bases from solution, principally chloride and sulfate. However, unlike WAC resins, WBA resins remove the weak bases as their acid conjugates, e.g., H_2SO_4 and HCl. WBA resins are efficiently regenerated, and may be useful ahead of SBA exchangers if the water contains high levels of chlorides or sulfates.

WBA exchangers are also sometimes used ahead of SBA exchangers if the water contains more than 1 or 2 ppm of natural organics. Decaying vegetation produces complex organics that will foul SBA beads. WBA resins will remove these organics. The organics regenerate more easily from WBA resins than SBA resins.

Demineralizer Configurations and Mixed-Bed Exchangers

Table 3–5 illustrates various demineralizer vessel configurations. The reader will note the mixed-bed (MB) demineralizer mentioned in several of the options. Mixed-bed demineralization is the process that improved ion exchange performance so that it became suitable for ultra-pure applications.

Water from a cation/anion system, although of high quality, still contains too many contaminants for use in high-pressure boilers or as ultra-pure water at semiconductor manufacturing plants. A mixed-bed exchang-

Common Demineralizer Configurations

SAC/SBA	SAC = Strong Acid Cation
SAC/WBA/SBA	SBA = Strong Base Anion
WAC/SAC/WBA/SBA	WAC = Weak Acid Cation
SAC/SBA/SAC	WBA = Weak Base Anion
SAC/SBA/MB	MB = Mixed Bed

Table 3–5 *Common Demineralizer Configurations*

er contains both SAC and SBA resins, intimately intermixed. The exchanger performs as if it consisted of millions of miniature cation/anion units. A MB exchanger usually serves as a polisher of effluent from a cation/anion system, reverse osmosis unit, or other purification arrangment. The mixed-bed will reduce contaminants from ppm levels to ppb levels.

Because a MB exchanger has such a light ionic load, flow rates can be increased. A typical cross-sectional flow rate is 20 gpm/ft^2, but higher flow rates are sometimes possible. The mixed bed is sized to provide long term operation with infrequent regenerations. This can be done without excessive quantities of resin, because the ion loading is so low.

Of the systems outlined in Table 3–5, the SAC/SBA/MB arrangement has been most popular, especially for high-purity applications including feed to boilers that operate above 1000 to 1500 psig. Selection of the best makeup system for lower-pressure units is a bit more complex. A system with a MB exchanger would obviously provide water suitable for industrial and small utility boilers, but this represents excessive treatment. At pressures below 600 psi, sodium softening is sometimes adequate for makeup treatment. The softener contains SAC resin in the sodium not hydrogen form, as only hardness ions need be removed from the water. The great advantage of this process is that the resin can be regenerated with a brine solution rather than more expensive acid.

Where mixed-bed quality water is required, but the plant operators and managers wish to avoid the hassles of mixed-bed operation and regeneration, bottled polishing is an option. A number of water treatment companies contract out this service, in which they bring in mixed-bed resin in portable containers, which can be quickly hooked up to the treatment system. The company takes away the used containers and regenerates the resin off-site.

Degasifiers

Alkalinity that passes through a cation bed is converted to carbonic acid (H_2CO_3), which is essentially just hydrated carbon dioxide ($CO_2 \cdot H_2O$). If the carbon dioxide is allowed to enter an anion exchanger, the CO_2 will convert to carbonate and increase the load on the anion unit. This load can be reduced if the CO_2 is removed upstream. Forced-draft or vacuum degasification will accomplish this. In the former mechanism, water is cascaded down a series of trays or splash plates while air is blown upward (Figure 3–15). Forced-draft decarbonation will reduce carbon dioxide concentrations to approximately 10 ppm although it does saturate the water with oxygen. A vacuum degasifer pulls all gases from the tower as the water flows down the trays. Even lower CO_2-removal efficiencies may be obtained, but a vacuum degasifier is more expensive. The cost of a degasifier must be weighed against the value obtained by decreasing the load on the anion exchanger. Caustic prices generally tend to be high, so a degasification system may be warranted for even moderately alkaline waters.

Regeneration and Co-current/Countercurrent Systems

One of the major factors that influences demineralizer performance is regeneration, and how efficiently and well the resin can be restored to optimum capacity after it becomes exhausted.

Consider again Figure 3–14, which outlines the behavior of cation resin during operation. As we have seen, calcium is preferentially removed, followed by magnesium and then sodium. Eventually the available capacity of the resin becomes exhausted, upon which sodium begins to break through. At this point, regeneration is necessary. In the United States, sulfuric acid is the typical regenerant, while in Europe many systems are designed to use hydrochloric acid. The chemical process is similar in both. Although Equation 3–11 indicates that the exchange process is an equilibrium reaction, the reaction is driven to the right due to the higher affinity for cations over hydrogen ions. This is what makes the process so effective, and why ion exchange systems can produce water with just a few parts-per-billion of impurities. The natural question becomes, "If cations have a higher affinity, how can the resin be regenerated?" The answer lies in LeChatelier's Principle. By introducing a strong acid concentration to the resin, the abundance of hydrogen ions will replace many of the cations. A common average sulfuric acid regenerant concentration for SAC resins is 4%. Think about the difference between this solution and the water being treated. The influent water may have a total dissolved concentration of several hundred ppm. Four percent H_2SO_4 is the same as 40,000 ppm! The much greater concentration of acid forces Equation 3–11 to the left.

Fundamentals of Steam Generation Chemistry

Even with this effect, many cations still cling to the resin, and a point is reached where ecomonics dictate the extent of regeneration. A typical regeneration process may only restore 50% to 60% of the resin capacity, so the system must be designed to take this into account. Efficiency of regeneration leads us into the topics of co-current and countercurrent regeneration techniques, which we will examine in the next section. First is a point not to be overlooked about SAC resins.

Although 4% is a common recommended concentration for sulfuric acid regenerant, a step-wise increase from say 2% to 4% to perhaps 6% or 8% may be necessary. The low initial concentration prevents precipitation of calcium sulfate within the bed. After a suitable period of time, when some of the calcium ions have been removed, the regenerant concentration can be increased. I have worked at facilities in which both methods were used. At

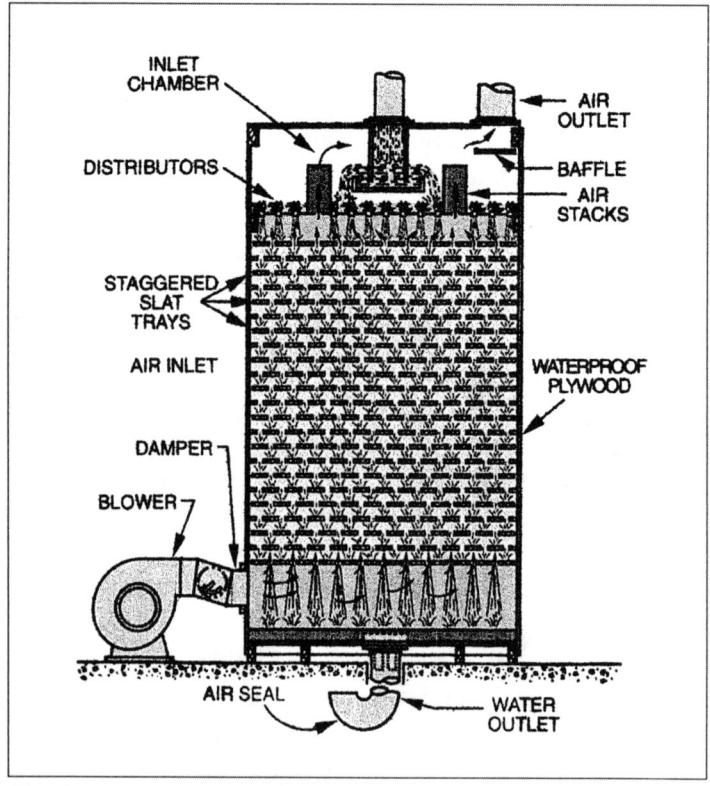

Fig. 3–16 *Forced Draft Aereator*
Source: *Betz Handbook of Industrial Water Conditioning, Ninth Edition. Betz Dearborn is a division of Hercules, Inc.*

the first, a straight 4% concentration worked well, because the raw water (from a surface supply) was relatively soft. In the second case, the supply came from wells, and calcium concentrations were rather high. The regeneration sequence actually called for a 1%, 2%, and 3% step-wise pattern.

A significant influence on regeneration efficiency is co-current vs. countercurrent regeneration. First generation demineralizers were designed as co-current units (Figure 3–16), in which the regenerant is introduced in the same path as the service water. While this is the simplest regeneration arrangement, it is not the most efficient. When regenerant is introduced co-currently, the most strongly held ions must flow completely through the bed before they are eluted. Thus, the cations can attach and then detach from the exchange sites as they pass through. This requires extra regenerant to force the cations out of the bed, and also leaves some cations at the bottom of the bed where they will show up in the effluent during the next service run.

One of the first techniques developed to alleviate this problem was air mixing of the resin. Although not an ideal solution, it did improve regeneration efficiency by destratifying the calcium and magnesium layers. The most dramatic improvement came with the introduction of countercurrent regeneration. In this process, the regenerant is introduced in the reverse direction of the service flow (Figure 3–17). It is important that the bed remain intact during regeneration. Several methods were developed to accomplish this, with two becoming most popular:

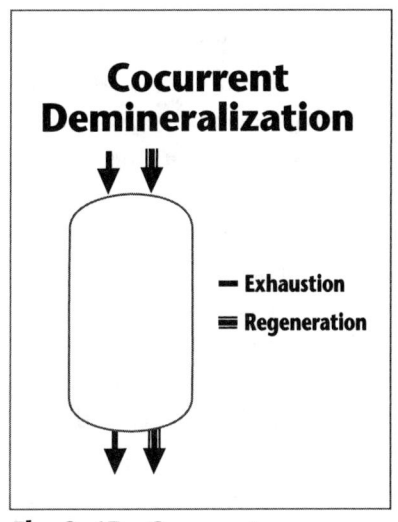

Fig. 3–17 *Cocurrent Demineralization*
Courtesy of the Purolite Co., Bala Cynwyd, PA.

- Introduce the regenerant at the bottom of the vessel and use a blocking flow of water at the top to keep the bed in place.

- Introduce the influent at the bottom of the vessel and inject regenerant from the top.

Countercurrent regeneration offers a very obvious advantage; the most tightly held ions are eluted without having to pass through the entire bed. Countercurrent regeneration can reduce chemical requirements, but perhaps more importantly allows a better quality water to be produced.

Strong Base Anion Regeneration

The performance and regeneration of a SBA exchanger is very similar to the SAC process mentioned above. When the resin reaches complete exhaustion, silica is the first ion to break through. SAC beds are regenerated with sodium hydroxide, usually in a 4% concentration. Stepwise regeneration of an anion bed is not usual because the caustic does not form a precipitate with any of the anions. As with cation exchangers, countercurrent regeneration is more efficient than co-current regeneration. Silica removal is greatly enhanced by heating the regenerant water. For Type I resins, the optimum temperature is 120°F. Type II resins are not as thermally stable and should not be heated above 105°F.

Figures 3–18 and 3–19 show the general arrangement for acid and caustic regeneration systems, respectively. The regenerant and dilution water are blended at a mixing station and then pumped to the exchange vessel. Nowadays, these systems are almost always automated. Two methods exist for measuring and controlling the acid or caustic concentration. Where demineralized water is used for dilution, the conductivity of the dilute solution correlates very closely to the acid or caustic concentration. Thus, on-line conductivity can be used as a monitoring and control device. A second method to check concentration is through manual measurement of the fluid specific gravity with a hydrometer.

Proper regenerations require high-grade regenerant chemicals so as not to contaminate the resin during the process. Learning Aide 3–3 outlines specifications for sulfuric acid and caustic that meet these requirements.

Weak Acid and Weak Base Exchangers

The primary purpose for WAC and WBA exchangers is to reduce the load on downstream SAC and SBA resins. This is economical because of the superb regeneration efficiency of these resins. Whereas SAC and SBA

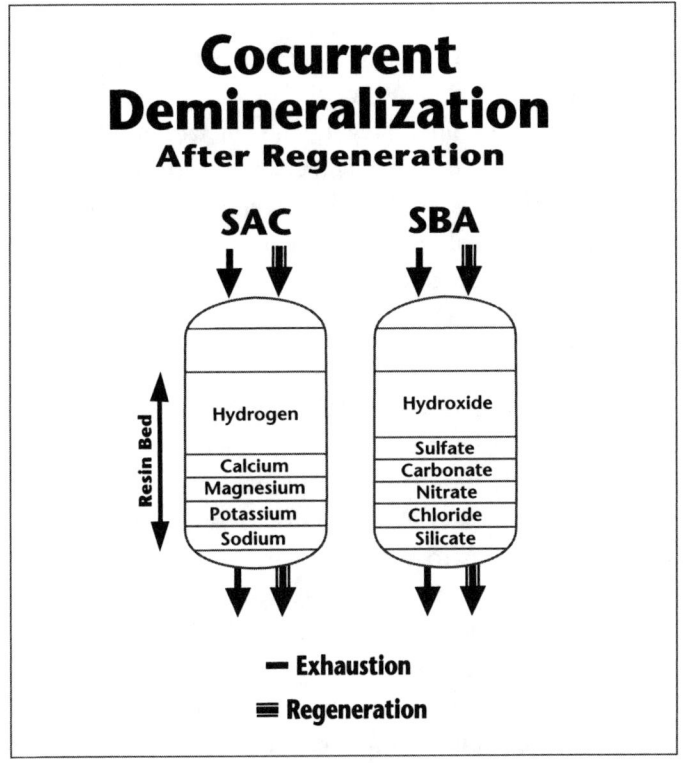

Fig. 3–18 *Cocurrent Demineralization after Regeneration*
Courtesy of the Purolite Co., Bala Cynwyd, PA.

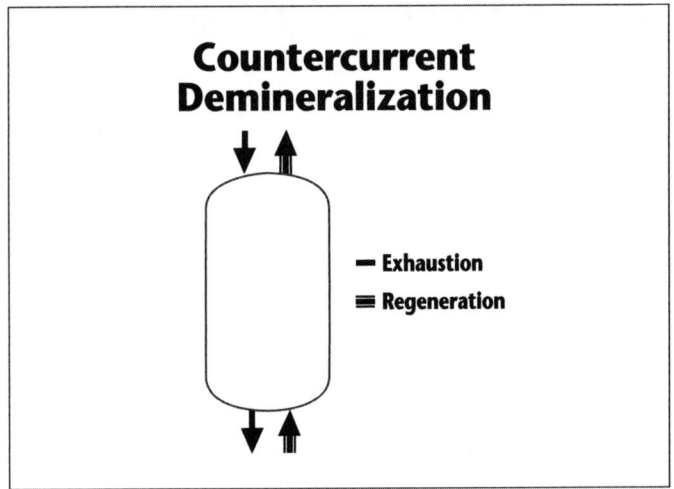

Fig. 3–19 *Countercurrent Demineralization*
Courtesy of the Purolite Co., Bala Cynwyd, PA.

resins may require regenerant with three times the hydrogen or hydroxide capacity of the ionic loading, weak beds can often be restored with near stoichiometric amounts of regenerant. In fact, in most systems utilizing weak acid or weak base exchangers, the waste regenerant from the SAC or SBA exchanger is used as the regenerant solution for the weak bed. This reduces chemical consumption over that which would be used to regenerate the beds separately.

Mixed-Beds

Although regeneration of mixed-beds is an infrequent process, it is very important and must be handled with care, since two resins are involved. Regeneration is usually performed in the exchange vessel. Anion resin is lighter than cation resin, and when the MB resin is backwashed, the process causes the two resins to settle into distinct layers. Figure 3–20 illustrates the generic outline of a mixed-bed exchanger. It also shows common demineralizer internals, including the central collection header found in mixed-bed vessels. The central collecter sits at the cation/anion resin interface. During regeneration, acid is introduced below the cation resin and flows upwards. Caustic is introduced above the anion resin and flows downwards. The waste regenerant from each is collected at the interface.

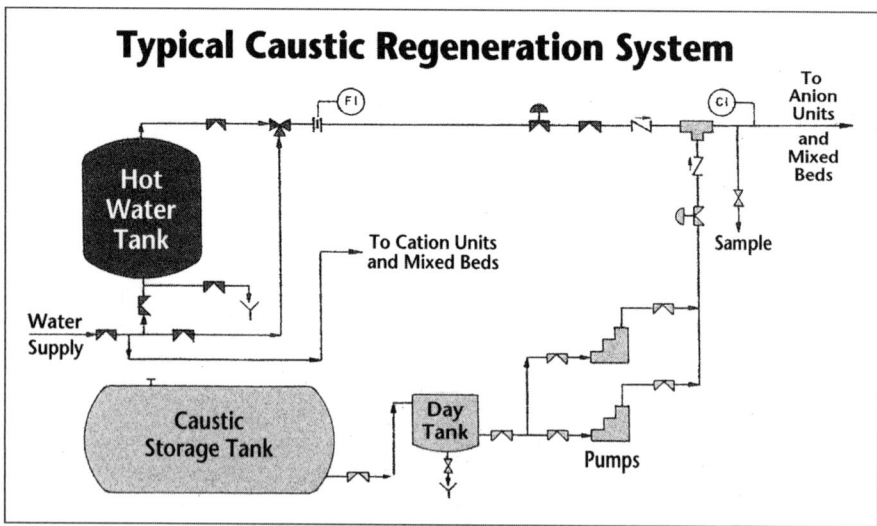

Fig. 3–20 *Typical Caustic Regeneration System*
Source: BetzDearborn, Inc., a division of Hercules, Inc.

CHAPTER 3: MAKEUP WATER TREATMENT

It is very important that the resins settle properly and that the division between them occurs at the central collector. If cation resin remains in the anion zone, it will absorb sodium from the caustic regenerant. Conversely, anion resin will pick up sulfate from the acid regenerant. A technique for providing good separation is to design the system with a small amount (10 percent or so) of inert resin that has a density between that of the cation and anion resin. The inert resin will settle at the interface, and provide separation between the active resins.

The mixed-bed regeneration process is similar to that for cation and anion resins with regard to acid and caustic concentrations. Once the resin has been rinsed, it is remixed with air. The air remix process is very important to re-establish the heterogeniety of the cation and anion resins. Typical guidelines suggest an air flow rate of 7 to 10 scfm for 15 minutes.

Monitoring Performance of Ion Exchanger Vessels

Figure 3–21 illustrates the typical effluent water quality from a cation bed during a normal service cycle. At startup of a cycle, the effluent is dumped to waste because it contains trace amounts of acid. The acid residual usually disappears quickly. The effluent water can then be routed through the rest of the system. During the service run, the cation effluent contains the free mineral acids H_2SO_4, HCl, and also H_2CO_3 and

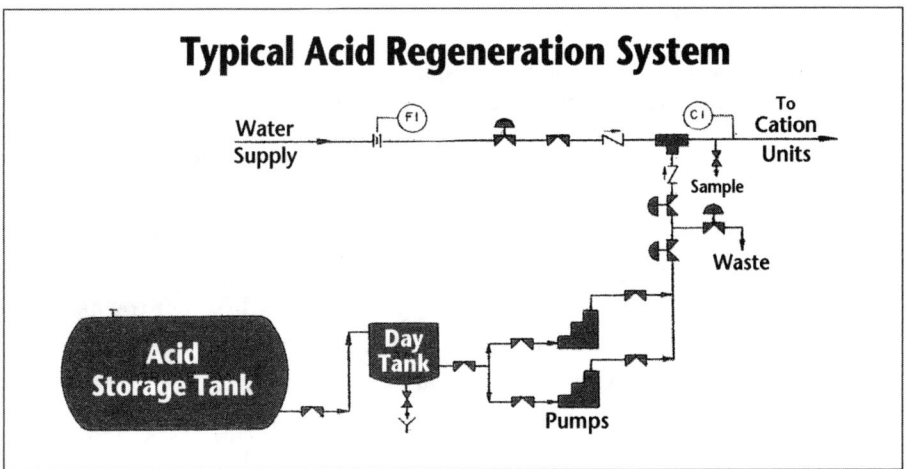

Fig. 3–21 *Typical Acid Regeneration System*
Source: BetzDearborn, Inc., a division of Hercules, Inc.

a small amount of sodium. Once the bed has reached exhaustion, sodium begins to break through. The free mineral acidity (FMA) and conductivity both begin to decline because sodium starts to take the place of hydrogen ions. On-line sodium monitoring can be very effective in detecting cation resin exhaustion.

The quality of water during an anion service run is outlined in Figure 3–22. A slight conductance is always present in the effluent, because the sodium ions that leak from the cation exchanger form sodium hydroxide in the anion bed. The conductivity usually remains constant throughout

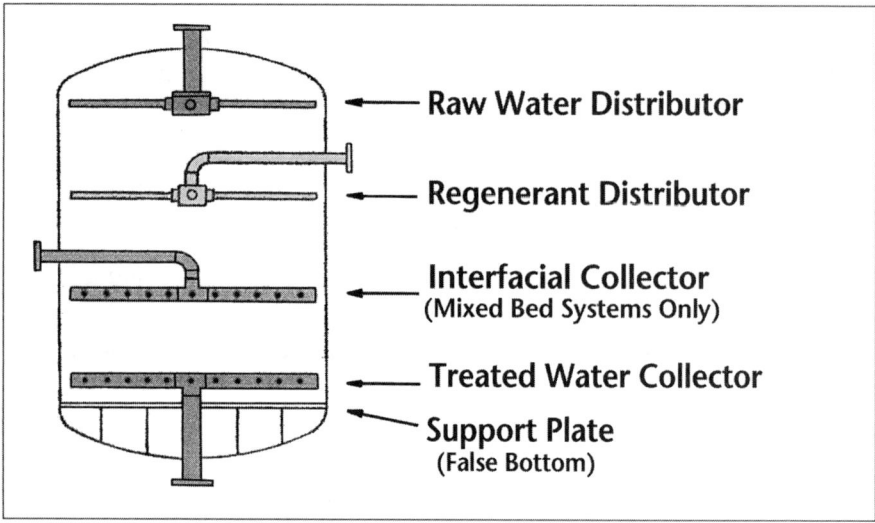

Fig. 3–22 *Ion Exchange Vessel Internals*
Source: BetzDearborn, Inc., a division of Hercules, Inc.

the service run. Likewise, the effluent contains a trace amount of silica due to leakage of this weakly-held constituent. When the anion exchanger exhausts, silica levels begin almost immediately to rise. On-line silica monitoring can be a very effective tool to detect resin exhaustion. Another phenomenon also occurs that can be used as a tool. When the bed just begins to exhaust, the first bit of silica that comes off combines with sodium to form sodium silicate. This compound has a lower conductivity than NaOH. With good observation, the small dip in conductivity can be used to indicate bed exhaustion. Because this measurement requires a trained eye, on-line silica monitoring is a better choice for most facilities.

CHAPTER 3: MAKEUP WATER TREATMENT

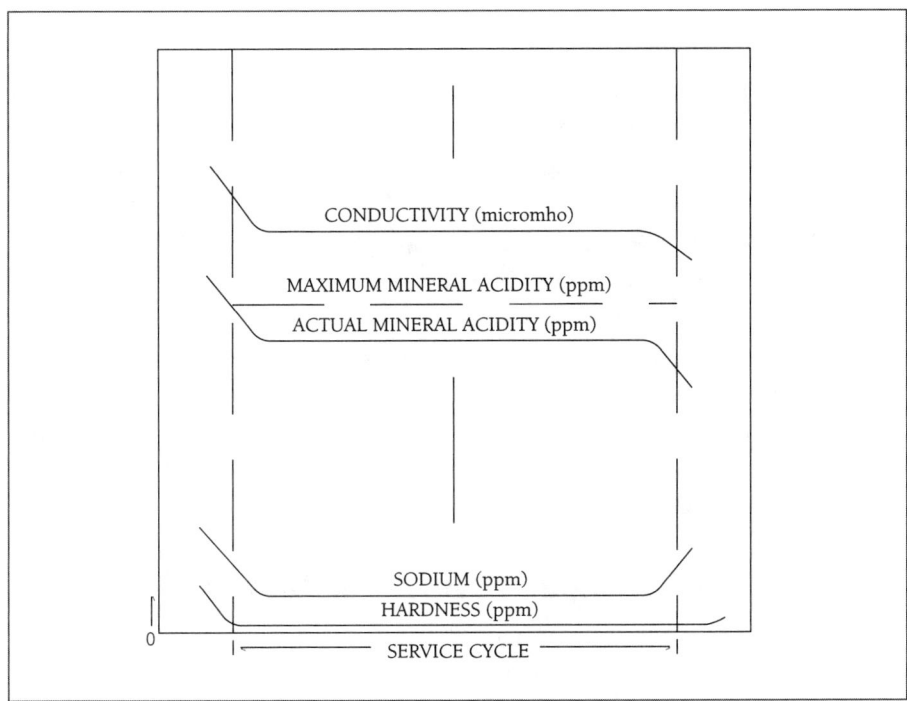

Fig. 3–23 *Water Quality from a SAC Exchanger during a Normal Service Run*

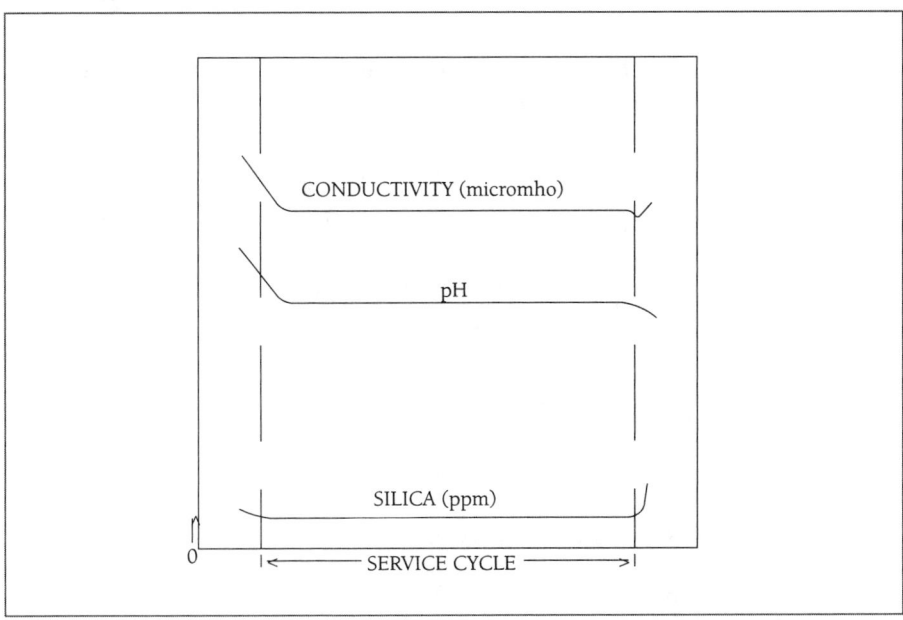

Fig. 3–24 *Water Quality from a SBA Exchanger during Normal Operation*

For systems with a mixed-bed exchanger, the mixed-bed acts as a buffer, at least for a while in the event of a cation or anion exchanger overrun. However, if the mixed-bed fails or becomes exhausted, nothing prevents the contaminants from travelling to the steam generating system. On-line sodium and silica monitoring of the mixed-bed effluent is highly recommended.

Many plant operators run the demineralizers to a certain setpoint, and then shut the system down for regeneration. The throughput volume is less than the known point of bed exhaustion. This method has merits, because it can help prevent demineralizer overrun and system upsets. The primary drawback is that more regenerant chemical is needed, and more frequent regenerations are harder on the resin. Even when control of the demineralizer is based on throughput, on-line monitoring of the effluent it still recommended.

Troubleshooting of Demineralizers

Even though demineralizers may run efficiently and reliably for many years, from time-to-time problems may crop up. These may be related to water chemistry, resin aging, or mechanical malfunctions. The following by no means exhaustive list illustrates some of the problems which can occur. As a preface to the list of mechanical failures, please refer to Figures 3–25, 3–26, and 3–27. These show the general outline of internal distributors and system face piping.

- Poor Quality of Effluent — Resin fouled, resin beads broken due to aging or stress, resin chemical breakdown, resin channeling, inadequate regeneration, poor water distribution due to failed or plugged distributors, failed or leaking valves, resin loss, high flow rates, change in makeup chemistry.
- High-Pressure Drop — Most of the above, low makeup temperature.
- Low-Pressure Drop — Loss of resin, high makeup temperature.

Packed Bed Demineralizers

The latest generation of demineralizers include systems based on packed-bed technology. In these demineralizers, the vessel is almost com-

CHAPTER 3: MAKEUP WATER TREATMENT

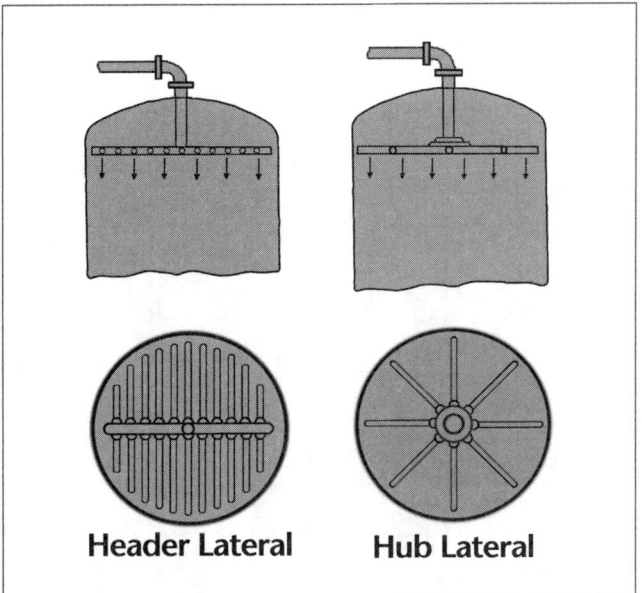

Fig. 3-25 *Common Distributor Arrangement I*

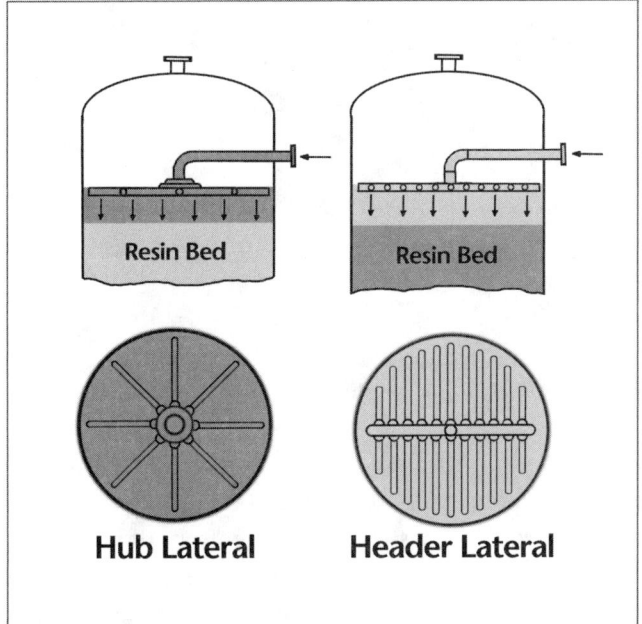

Fig. 3-26 *Common Distributor Arrangement II*

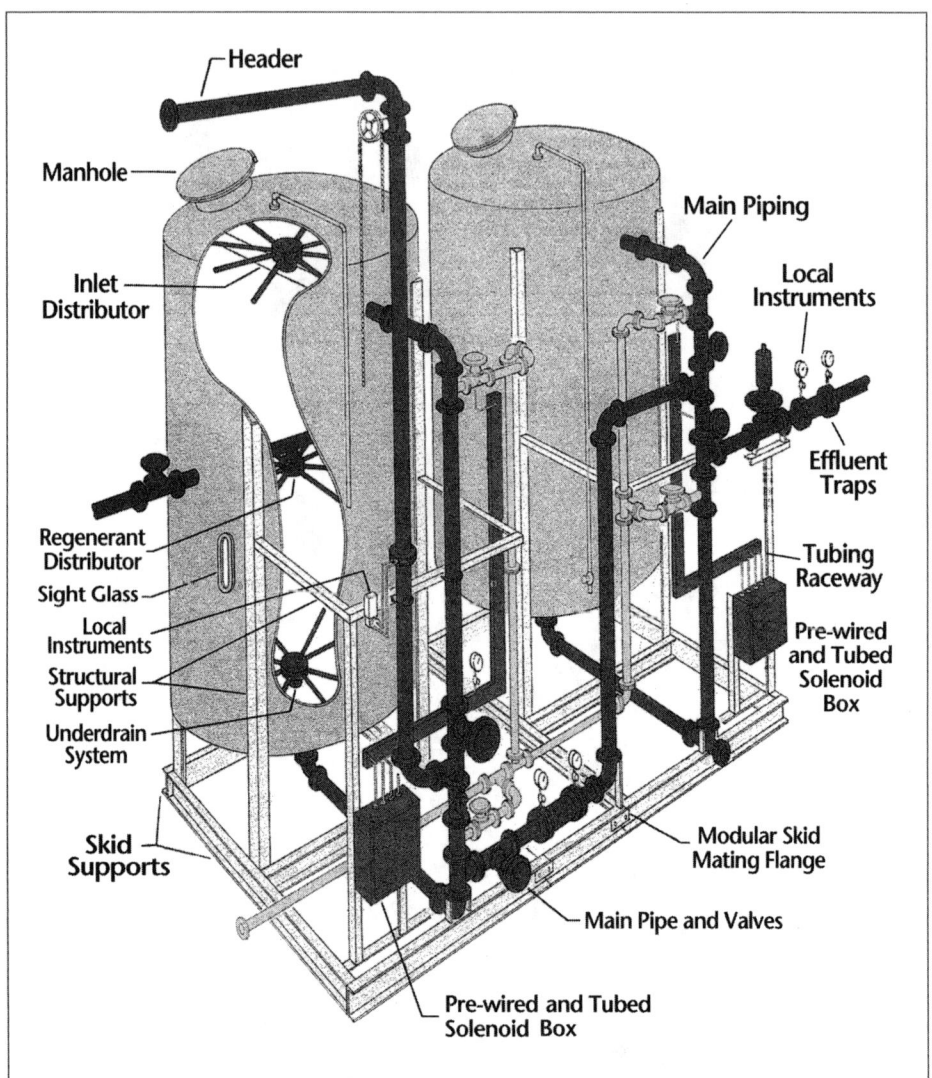

Fig. 3–27 *Outline of Piping on a Twin-bed Ion Exchange System*

pletely filled with resin, with perhaps only 10 percent freeboard. Two variations are most common, and both are based on countercurrent operation. In one, the process water flows from top to bottom of the vessel, and the regenerant from bottom to top. Unlike conventional countercurrent regeneration, however, the regenerant is introduced at enough pressure to lift the entire bed as a plug to the top of the vessel. The other type of packed-bed system essentially reverses these functions. The process water raises the bed, and the resin is regenerated from top to bottom. These processes require different distributor and collection systems than conventional units, whose distributor laterals would be subjected to mechanical stress by the moving plug of resin. Distributor and collecter plates with screened openings are located at the bottom and top of the packed-bed vessel, and serve as the restraining barrier for the resin.

Manufacturers of these systems claim several advantages versus regular countercurrent demineralizers. One advantage is that a smaller vessel will hold the same amount of resin as a standard ion exchange unit. Better regeneration efficiency is another reported advantage. This is in part due to the more intimate contact of regenerant with the tightly compacted bed than occurs in a standard countercurrent unit where the bed may be somewhat fluidized.

One area where packed-bed technology has been applied is retrofit of existing co-current demineralizers. Water quality can be significantly improved by this change.

Another interesting technology is the short-bed demineralizer. In these systems, the resin is truly packed into minimum-depth vessels. No freeboard exists. The main features of short-bed systems include:

- The bed height may be as short as 6 inches.
- Very fine resins are used for improved reaction kinetics.
- Surface area flow rates may be as high as 30 gpm/ft^2 as compared to standard design values of 4 to 12 gpm/ft^2.
- Run times range from 5 to 30 minutes with a 20 to 40 second regeneration and a one minute rinse.
- After regeneration the effluent is recirculated to the influent until water quality reaches desired values. This may take a couple of minutes.

The obvious advantage of this system is its small size. For facilities with limited space, or where a demineralizer must be retrofitted into a confined area, these systems can be very advantageous. Water quality and regeneration efficiency are also reported to be very good. One drawback to these systems is that with zero freeboard, the resins cannot be backwashed. Thus, the influent must be free of suspended solids.

Other Make-Up Technologies

Ion exchange does not have the same dominance it once had on the high-purity treatment market. This is principally due to the development of reverse osmosis technology. The hybrid ion exchange/membrane methods of electrodialysis and electrodeionization are also making inroads into the makeup water market.

Crossflow Filtration and Reverse Osmosis

Reverse osmosis is a member of the crossflow filtration family, which also includes microfiltration, ultrafiltration, and nanofiltration. In an ordinary depth or weave filter, water flows perpendicularly to the filter (Figure 3–28). Particles are removed throughout the depth of the filter. Over time, the filter plugs with solids and must be replaced. In crossflow filtration systems, water flows parallel to the membrane surface. Applied pressure at the influent forces water through the membrane as it passes from one end to the other. Solids are swept along with this flow and exit at the end of the pressure vessel.

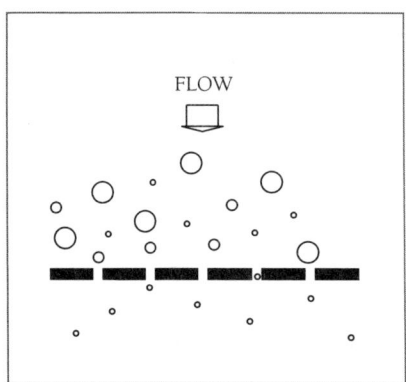

Fig. 3–28 *Conventional Filtration*

Figure 3-29 shows the filtration spectrum for the various membrane processes. For the time being, we will focus on reverse osmosis, as the discussion of RO provides useful information on membrane design, operation, and hazards. A subsequent discussion of MF, UF, and NF will be more clear.

The feature that makes RO unique as compared to other filtration methods is its ability to removed dissolved solids down to the smallest

CHAPTER 3: MAKEUP WATER TREATMENT

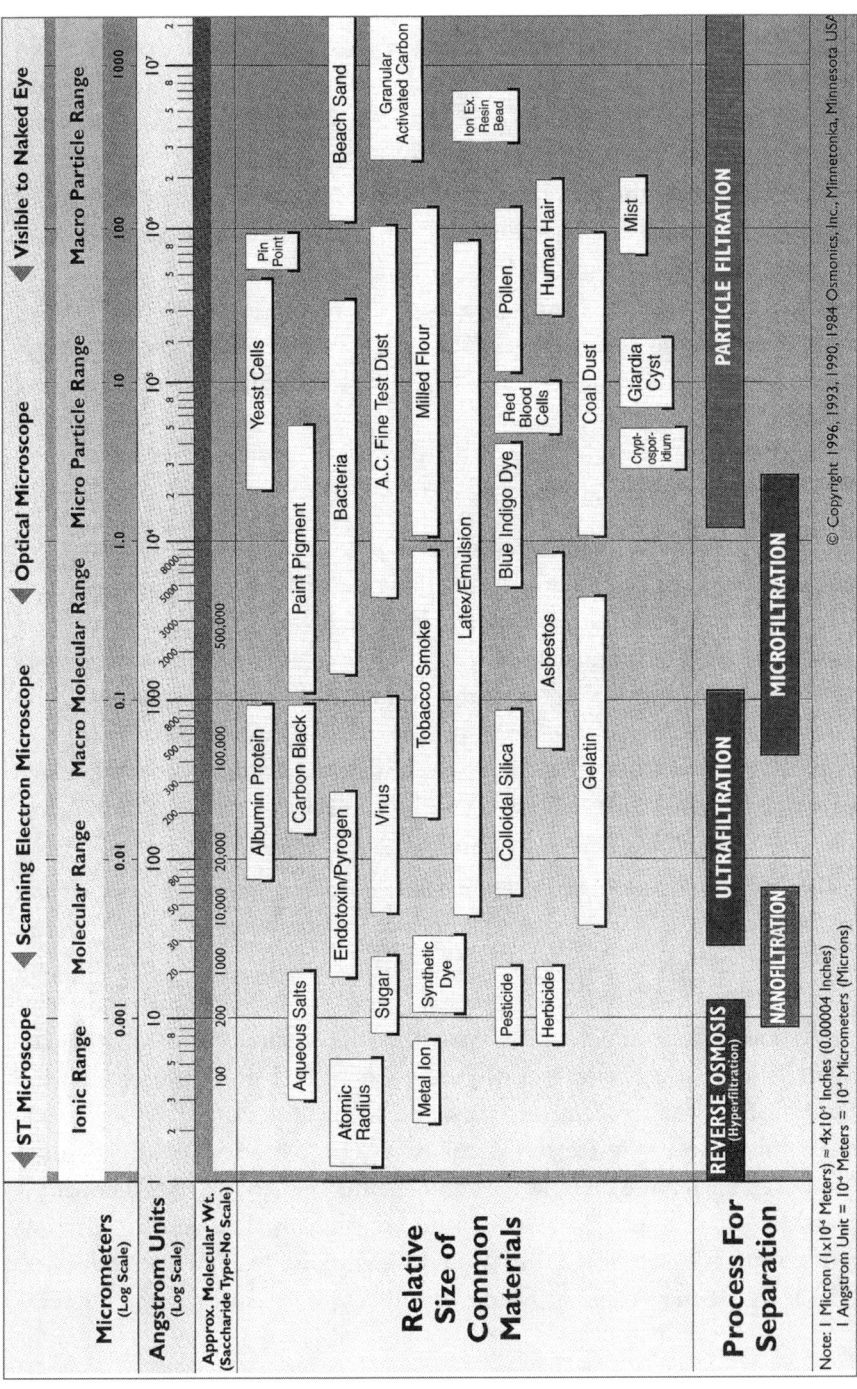

Fig. 3–29 *Filtration Spectrum* Courtesy of Osmonics, Inc., Minnetonka, MN

ions. This provides an advantage over ion exchange, as Practical Example 3–4 will illustrate. Although the passages through RO membranes are often referred to as pores, they more resemble very tiny maze-like tunnels. The passages range from 1 angstrom (10-8cm) to 10 angstroms in diameter. Because the pore diameters are so small, the molecular layer of water that attaches to the membrane surface inhibits ions from passing through the pores. This is unique to reverse osmosis, as all other filters including MF and UF remove particles by mechanical blockage. Even so, some ions and molecules may still find their way through the RO membrane. These include monovalent ions, such as sodium and chloride, and small organic molecules.

RO Membrane Design

Two types of RO membranes are most common today, hollow-fiber and spiral-wound. In hollow-fiber configurations, a membrane element consists of many fibers bundled togther. Each fiber individually treats a portion of the feedwater. The purified water (permeate) and concentrated water (reject) are collected at the end of the vessel. The fibers may be designed that water passes from inside-out or outside-in. We will look at these a bit more closely in the discussion on microfiltration. By far, spiral-wound membranes are more common for RO systems. The remainder of this section is based on spiral-wound systems.

Spiral-wound membranes are manufactured in flat sheets, which are wound around a central core to produce a membrane element (Figure 3–30). Several elements are placed in series and are sealed in a pressure vessel (Figure 3–31). Multiple pressure vessels in both parallel and series configuration make up a typical system. Within each pressure vessel, feedwater enters the forward end of each element and flows to the opposite end. Purified water (permeate) passes to the central core of the element, while the concentrate (reject) is collected and discharged at the element end cap. The feature of spiral wound membranes that make them most practical is the multiple wraps within an element. This lets an element process much more water than would be capable through a single sheet. One element may contain several hundred square feet of membrane.

CHAPTER 3: MAKEUP WATER TREATMENT

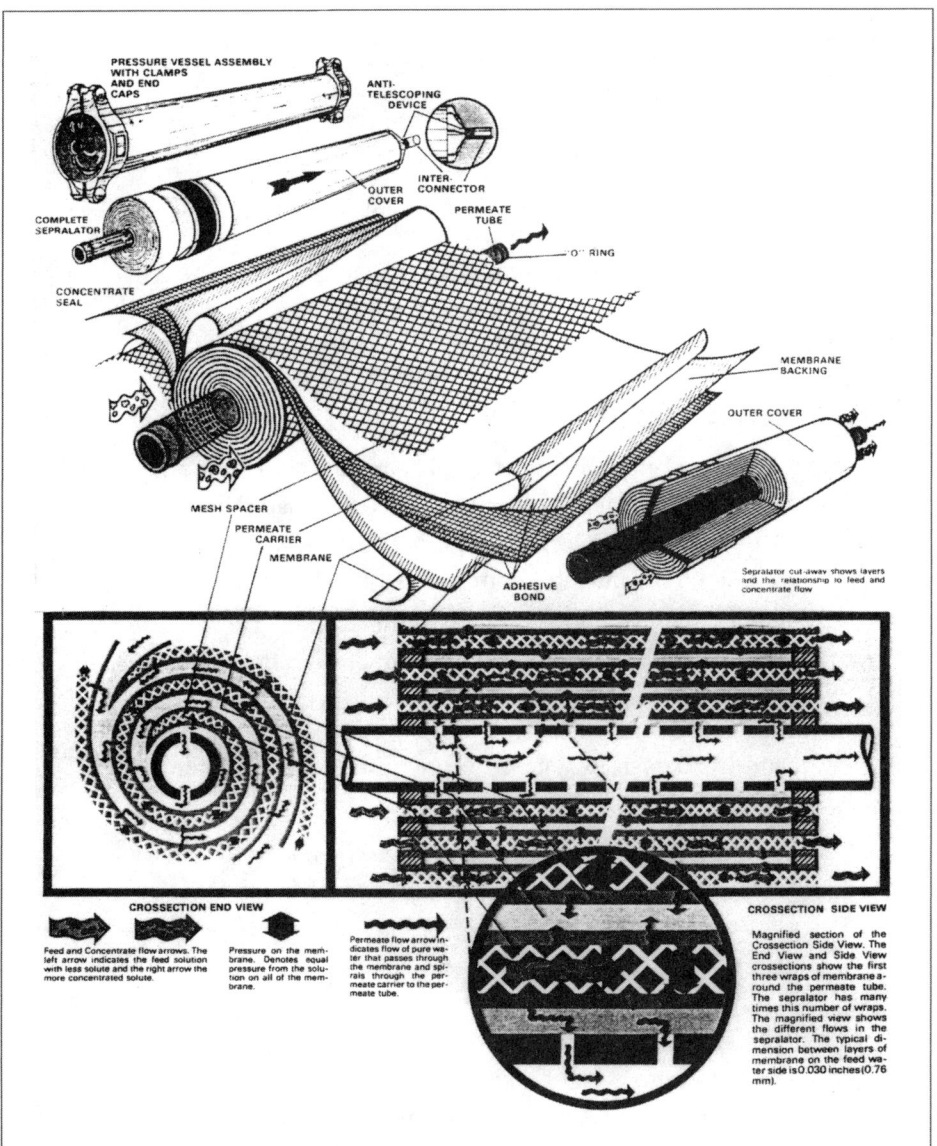

Fig. 3-30 *Spiral-wound Membran Structure*
Courtesy of Osmonics, Inc., Minnetonka, MN

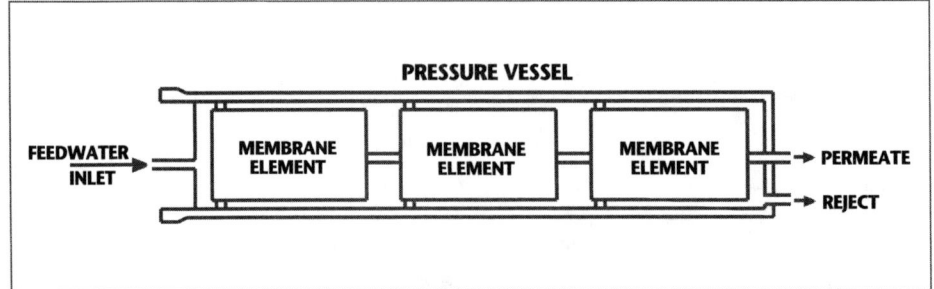

Fig. 3-31 *Generic Outline of a RO Pressure Vessel*
Courtesy of Osmonics, Inc., Minnetonka, MN

RO Membrane Material

Two types of membrane material are most common, cellulose acetate/triacetate (CA) and polyamide. The latter material is typically layered with other membranes for support. These layered polyamide membranes are known as thin-film-composites (TFC).

CA membranes were the first to be developed. The material offers both advantages and disadvantages. The important advantages include:

- The membrane is chlorine tolerant up to about 1.0 ppm and works well in systems where microbiological fouling must be controlled.

- The membrane is nonionic and has less tendency to attract ionically charged substances such as coagulants and water treatment polymers.

Disadvantages include:

- Feed pressure is higher than for TFC membranes. Typical feed pressures of CA systems for the applications described in this book range from 200 to 400 psig.

- In the absence of a microbiocide, CA membranes will be attacked and degraded by microbes.

- CA membranes are only stable within a narrow pH range of 4 to 6. Excursions outside of this range will cause membrane degradation. Since many waters are neutral to alkaline, this may require acid feed ahead of the system.

TFC membranes also have advantages and disadvantages. The advantages include:

- Operating pressures are lower than with CA membranes. TFC membranes have been developed that operate very well with feed pressures as low as 150 psig. Some reports indicate that feed pressures may go as low as 100 psig.
- TFC membranes are not attacked by microbes. However, they can be fouled with microbiological deposits.
- TFC membranes are stable over a pH range of 2 to 12.

Disadvantages include:

- The membranes are attacked by oxidants. This requires activated carbon filtration or an antioxidant feed ahead of the system.
- The membrane surface exhibits a negative charge, which will attract coagulants and cationic polymers.

RO Pretreatment

Good pretreatment is a critical issue for reverse osmosis membranes. They can be easily fouled by suspended solids. Also, the concentrating effect as water passes along the membranes can lead to scale formation. As we have seen, TFC membranes will not tolerate oxidizing biocides while CA membranes must operate in a narrow pH range.

The makeup water pretreatment methods outlined earlier in this chapter will often produce water suitable enough for feed to a RO. A rule-of-thumb says that waters with a turbidity of less than 1 NTU are acceptable. This is only a general guideline. Another measurement for determining the fouling potential of water is known as the Silt Density Index (SDI). SDI is a measure of the effect suspended solids have on water flowing through a filter. Learning Aide 3–4 outlines the SDI procedure, calculations, and guidelines. Usually, 10- and/or 5-micron depth filters are placed ahead of the RO to further minimize the potential for particulate fouling. Even these may not be a total cure, as Practical Example 3–5 illustrates.

Potential scale formation is another factor that must be taken into account. When water flows through a reverse osmosis pressure vessel, the concentrate continually accumulates dissolved solids. As we shall see, common RO system designs have one set of pressure vessels treating concentrate from another set. This increases the scaling potential. Calcium carbonate, sulfate, or other compounds can build up to a point where precipitation begins to occur. Other potential scales include silica and alkaline metal silicates, strontium sulfate, barium sulfate, and calcium fluoride. While pretreatment may reduce the concentrations of many scale forming compounds, the remainder may still cause problems. Barium and strontium sulfate scales are especially difficult to deal with. The reputable membrane manufacturers have developed programs that will calculate the solubility limits for these salts in a particular application. The program will warn the user if any solubility limit is exceeded.

Where scaling is a potential problem, antiscalants can help. Common antiscalants include polyacrylates and phosponates. The correct antiscalant or blend can control calcium sulfate at 230 percent above the saturation limit, strontium sulfate 800 percent above the saturation limit, and barium sulfate 6000 percent above the saturation limit.

Water treatment chemicals can also affect membrane performance. Coagulating agents of the cationic variety are particularly troublesome to RO membranes, especially to TFC membranes whose surface is negatively charged. This has been an overlooked item. If these agents are present, methods to remove them must be considered. More troublesome may be microbiological fouling. Microbes will attack CA membranes, and they will foul TFC membranes. Practical Example 3–6 outlines a case history of such fouling.

For those who may eventually be tasked with specifying, purchasing, or installing a RO system, the importance of obtaining accurate influent water quality data cannot be overemphasized. Ideally, historical data will be available. If not, then analyses should be collected as far in advance as possible of the decision to purchase a reverse osmosis unit. I made the mistake of using a "snapshot" SDI analysis to affirm that a reverse osmosis unit could be placed ahead of a demineralizer at an electric utility. The analysis, conducted by the eventual RO supplier, showed an SDI of less than one. After the RO was purchased, plant personnel rechecked the SDI and obtained some readings above 5. The fluctuations were due to changes in the surface water supply and performance of a Lamella clarifier.

RO Design

A reverse osmosis unit has been called nothing more than a high-pressure pump, some pressure vessels, and pipe. In truth, the operation is more complicated than this. Spiral-wound membrane elements can come in several different sizes. The most popular size is 8 inches in diameter by 40 inches in length. These are loaded in series into a pressure vessel (Figures 3–30 and 3–31), with four, five, or six elements per vessel being most common. Each element can pass a certain amount of water, and this volume is usually measured in gallons per day (gpd). Common values for 8" x 40" elements range from 4000 to over 13,000 gpd. The rate at which water passes through the membrane is known as the flux and is measured in gallons per square foot per day (GFD). The general purity of the water partially dictates the flux rate. A general guidelines suggests:

- Surface Water — 8 to 14 GFD
- Well Water — 14 to 18 GFD

For normal surface and groundwaters, each pressure vessel will produce about 50 percent purified water (permeate) and 50 percent concentrated water (reject or concentrate). This does not seem very efficient. However, the concentrate is often still pure enough to be treated again at another 50/50 split to produce 75 percent capacity. Figure 3–32 illustrates the generic design of this two-stage RO system. Sometimes even the second concentrate can be treated to give an overall RO output of 87.5 percent.

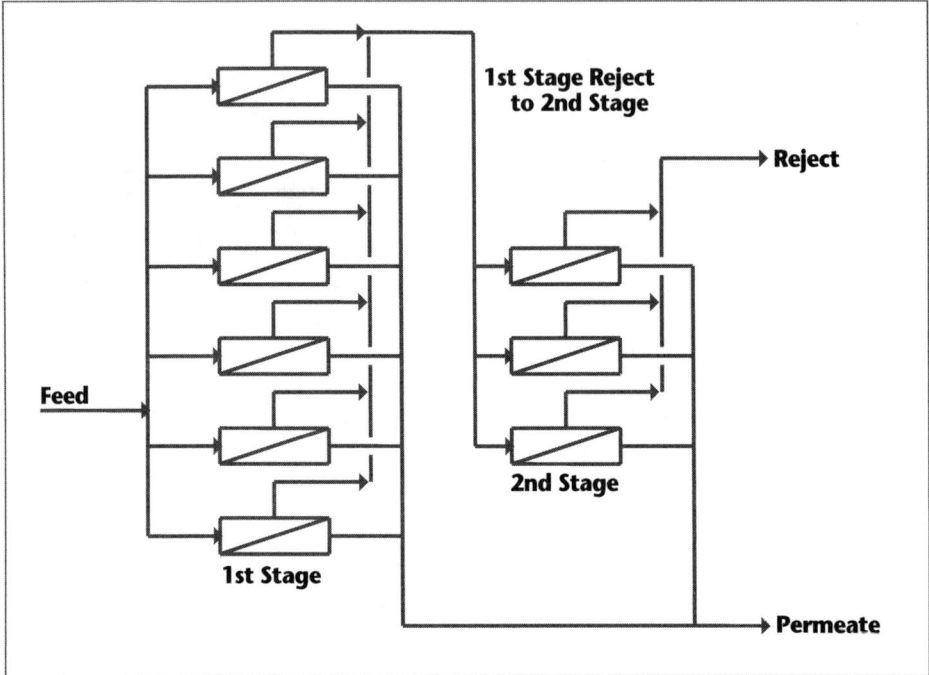

Fig. 3–32 *Two-stage RO Schematic*

To illustrate the calculations that go into sizing of a RO system, let's look at a unit designed to produce 300 gpm of purified water with 75 percent recovery of the influent. Conditions are as follows:

- Surface Water
- SDI < 5
- Five Membranes per Pressure Vessel
- Membrane Capacity - 10,000 gpd

The system will be a two-stage process, whereby the concentrate from the first stage is sent to a second stage for further treatment. Since 300 gpm of purified water is to be produced, and the overall efficiency is 75 percent, the influent quantity must be 400 gpm. This is 576,000 gallons per day. At a capacity of 10,000 gallons per day per element, 58 first-stage elements are required. Rounding off gives 12 pressure vessels with 5 elements apiece in the first stage. Because 50%, or 288,000 gallons per day, of the first stage is concentrate, the theoretical number of elements required in the second stage is 29. Rounding off gives 6 pressure vessels with 5 elements apiece.

This is a simplified examination of RO sizing, but it does provide an example of the flexibility that is available. The number of elements can be varied per pressure vessel, and the quantity of pressure vessels themselves can be adjusted to obtain the desired flow rate.

RO elements that offer 99 percent or greater salt rejection are now available. Over time, however, membranes will degrade until, after two or three years of operation, salt rejection may be only 95 to 97 percent. Even so, the RO still performs a valuable service. A very common application is retrofit of an RO ahead of an existing demineralizer. This can greatly extend the demineralizer run lengths and cut down on regenerant costs and resin replacement frequency. It is not impossible for regenerant chemical costs to exceed $100,000 per year for even a moderately-sized (200 gpm) demineralizer. Operating costs for the RO pump (power), perhaps a small amount of antiscalant feed, periodic chemical cleaning of the RO membranes, and (reduced) regenerant chemical feed to the demineralizer may be less than a fourth of the normal chemical regenerant cost. Payback time for the RO might be as short as two years. One factor not included is the cost for membrane replacement. Membranes typically last from 3 to 7 years, although longer life is possible. Membrane replacement costs may be a third of the original price of the RO.

A technique that is gaining acceptance for new high-purity water installations is reverse osmosis followed by mixed-bed polishing. This arrangement performs very similarly to an SAC/SBA/MB demineralizer. One advantage that RO provides over standard demineralization is its ability to remove collodial silica.

Various articles and technical papers have been presented over the last few years that attempt to give a rule-of-thumb guideline for the water quality at which RO is favored over SAC/SBA, when installed ahead of a mixed-bed. Although the topic is debatable, it is evident that the economics have improved for RO. Around 1990, it was suggested that an influent TDS of approximately 350 ppm or higher made RO economical. Reports later in the decade indicated that this level had dropped to as low as 150 ppm. I have since seen reports that suggest 100 ppm. However, these calculations are strongly influenced influent water quality, and must be examined on a case-by-case basis.

Yet another RO arrangement is the double-pass system, which should not be confused with the two-stage design. In this system, the permeate from the first stage is sent to another set of membranes for further purification (Figure 3–33). This process is capable of reducing dissolved solids concentrations to less than 1 ppm. The concentrate from the second pass is recirculated to the influent, so the overall efficency of the reverse osmosis system is not diminished by adding this second pass. Two-pass RO permeate is potentially suitable for feed to low- and medium-pressure boilers.

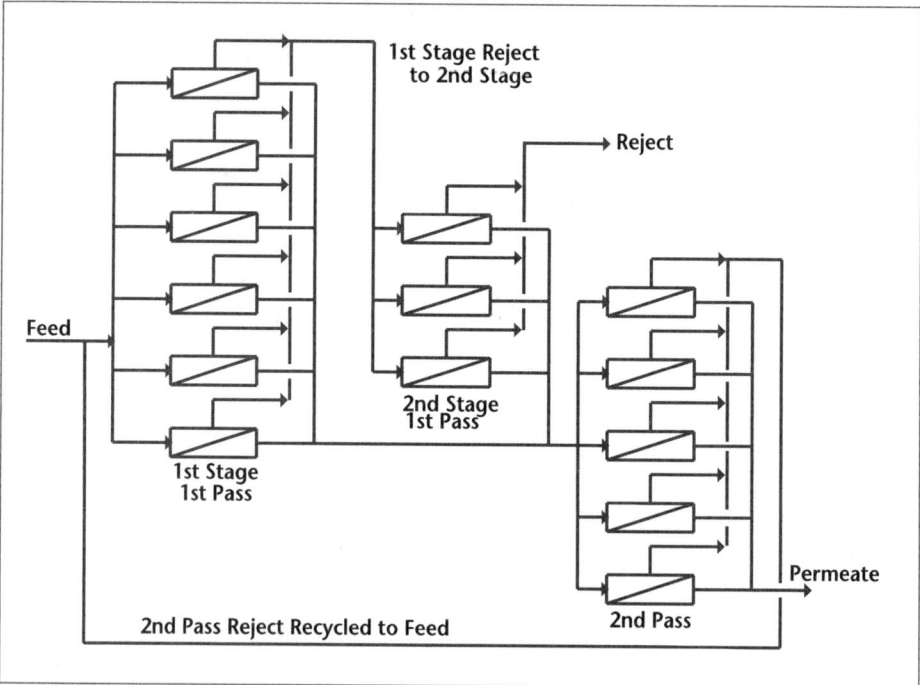

Fig. 3–33 *Two-pass RO Schematic*

CHAPTER 3: MAKEUP WATER TREATMENT

RO Components

An actual RO unit is shown in Figure 3–34. A number of components besides the membranes make up a reverse osmosis system. They include the pump, pressure vessels, and piping, and instruments. The following sections outline some of the most important aspects of this equipment.

Pumps. Centrifugal, multi-stage pumps have proven best for RO applications. They can easily generate the pressures needed, although pressure requirements have decreased as membrane quality has improved. Stainless steel is recommended as the pump material to give the pump long life and durability. RO feed pumps tend to be noisy, but submersible types are an option to minimize noise.

Pressure Vessels. Pressure vessels are typically fabricated from stainless steel or fiberglass. Both have proven effective. At one time, stainless steel housings were more flexible from a design standpoint because feedwater lines could be tapped into the side of the vessel. This arrangement allows a technician to remove membranes without having to disconnect any plumbing. Fiberglass did not have the structural strength to support side entry

Fig. 3–34 *Skid-mounted Reverse Osmosis Unit*
Photo courtesy of U.S. Filter.

lines. This has changed, and now fiberglass vessels are being constructed with side-entry ports. The vessels are designed with close tolerances, especially at the end caps so that the permeate and concentrate flow along the required paths and do not intermingle. Failure of O-rings or improper alignment of end-caps can lead to poor performance from a vessel.

Piping. For high-pressure RO piping, stainless steel is the popular choice. Often, however, one will see high-pressure flexible hose connecting one pressure vessel to another.

RO Flow Control and Monitoring.

The three most important measurements for monitoring RO operation are flow, pressure, and conductivity. Other important measurements include temperature, pH (especially for CA membranes) and oxidation-reduction potential (ORP, especially for TFC membranes). Flow is extremely important because it will change if membranes foul, tear, or degrade due to chemical or microbiological attack. Monitoring of flow rate, or perhaps more accurately, flow performance, is not a mindless task, as temperature directly affects the amount of water that will pass through the membranes. When water cools it becomes more dense, and its passage is restricted through the membranes. At the same pump pressure, flow may decrease by almost 50 percent when the influent temperature drops from 77°F to 50°F. Conversely, the flow may increase by 10 percent with a 10-degree rise in temperature above 77°F. A conversion calculation must be included in any flow measurements at temperatures other than 77°. This is called "normalizing" the flow rate. Without this correction, it would be difficult to accurately determine whether a change in flow rate was related to temperature or a system or membrane upset. Membrane manufacturers supply normalizing calculations that allow the operator to accurately normalize flow rates.

The change in flow due to temperature brings up an interesting problem when designing a reverse osmosis system. How should the feed pump and membrane quantity be determined? Several methods are possible. The system could be sized to produce the required capacity at the coldest inlet water temperature. The system might then be cycled more fre-

CHAPTER 3: MAKEUP WATER TREATMENT

quently in the spring and summer due to the increased capacity. This can place added stress on the unit. If an absolute flow rate is not required, the system could be sized for an average water temperature, with the knowledge that output would be greater or smaller depending on the time of year. A third option is to select a unit that provides the required flow at the coldest temperatures and use a variable speed feed pump to regulate flow at other times. A fourth possibility is to install a heat exchanger on the RO influent and regulate the influent temperature. This is a very viable option where auxiliary steam is available for heating.

Continuous conductivity monitoring of the permeate is very important. While membrane aging will cause a gradual increase in conductivity, a major problem such as a tear in a membrane, failure of an O-ring, or degradation of membranes by chemicals in the water or microbiological fouling, may cause a rapid degradation of effluent quality. Any such upset must be detected promptly. Where the problem is isolated to one pressure vessel, it may be necessary to manually check the conductivity from each vessel until the problem is identified.

With CA membranes, continuous measurement of pH is critical. Since many waters have a pH outside of the 4 to 6 range that is optimum for CA membranes, acid feed is a standard pretreatment. Continuous pH monitoring will alert the operator to overfeed, underfeed, or failure of the acid treatment pump. With TFC membranes pH is not generally an issue but the presence of oxidizing biocides is. This is where on-line ORP monitoring is valuable. As we learned in chapter 1, oxidizing agents are eager to grab electrons from some source. This potential can be measured, and reliable ORP monitors are available to calculate the oxidizing (or reducing) potential of aqueous solutions. Where chemical dehalogenation is used, continuous ORP monitoring will allow the operator or automatic equipment to adjust the feed rate or react to a system failure.

The flow rate of the reject or concentrate stream of a reverse osmosis system is always controlled. As water flows along the membranes it carries solids with it. This prevents the buildup of dissolved solids along the membrane surface. The reject flow rate must not be allowed to drop too low, as otherwise fouling could occur. (This concentration of solids is known as the Beta factor, and is calculated by membrane projection programs supplied by various manufacturers. It will notify the user if the design criteria selected give a low reject flow rate.) A value commonly used for determining the reject flow rate is known as the concentrate-to-permeate ratio. It applies to the individual elements. Concentrate-to-permeate ratios generally range from 4:1 to 9:1 depending on the quality of the influent.

Alarms

A number of circumstances might occur that could cause damage to a reverse osmosis system, particularly the membranes. Therefore, RO systems are typically equipped with a number of alarms or automatic shutdown devices to protect the equipment. These include.

High Permeate Pressure

Membranes are designed to function with flow in only one direction. Backpressure applied to the membranes from the permeate discharge could unravel the membranes. A high permeate pressure alarm and automatic shut down controller will protect the system from this type of upset.

High Concentrate Pressure

Backpressure may be exerted on the membrane if the reject valve should somehow be closed or flow become otherwise constricted. A high concentrate pressure monitor and alarm helps protect the system. An alternate technique to serve this and another purpose is reject flow monitoring and low alarm. Low reject flow is serious, because a decrease in flow rate reduces the ability of the concentrate to sweep solids from the membrane surface.

High/Low Influent pH or High ORP

These alarms indicate a problem makeup quality and the chemical feed systems that treat the RO influent.

High Permeate Conductivity

A sudden increase in conductivity indicates performance problems with one or more of the membranes or pressure vessels.

Low/High Pressure RO Feed Pump Pressure

Low pressure indicates a problem with the pump. High pressure indicates a possible obstruction or flow problem in the RO system.

Size of a Reverse Osmosis System

The most common membrane elements are 8 inches in diameter by 40 inches long. Four, five, or six of these mounted in series in a pressure vessel can extend the vessel length to over 20 feet. It is pressure vessel length that most determines the size of the system. The cylindrical vessels can be placed in a variety of parallel configurations, depending on the space requirement. They can be stacked or placed side by side.

When installing a reverse osmosis system in a building, ideally space should be made available at one or both ends of the system so that the membranes can be pulled for maintenance or replacement.

RO Cleaning

Periodically, membrane performance may decline enough to require a chemical cleaning. Because fouling is most prevalent on the concentrate side of the membrane, the chemical cleaning solution is injected through the reject line. The chemical is mixed up in an external tank is circulated with a low-pressure pump over the membranes. Usually, the cleaning system piping network is set up so that only a portion of the pressure vessels are cleaned at one time. The recommended flow rate for an 8-inch diameter pressure vessel is 40 gpm.

Common cleaning chemicals include citric acid for calcium carbonate removal, a carbonate/EDTA chelant for calcium sulfate, and alkaline phosphate/EDTA solutions for organically fouled resins. The strength of solution is usually between 1 and 2 percent. The solutions work better when warm, so the mixing tank is equipped with a heater to raise the temperature of the cleaning agent. 105°F is recommended.

For clean up of microbiological foulants, recommended treatments include a 10 ppm sodium hypochlorite solution for CA membranes and a 400 ppm peracetic acid solution for TFC membranes. From my own personal standpoint, I would be reluctant to perform this type of treatment unless absolutely necessary. I was a supervisor at the facility whose RO is outlined in Practical Examples 3–5 and 3–6. As a last resort, an equipment representative cleaned the membranes with peracetic acid. Membrane performance, which had not been great, rapidly declined after the cleaning and required replacement within weeks. An alternative is to clean with non-oxidizing biocides or feed a non-oxidizer ahead of the unit.

Practical Example 3-4

This occurred at my former utility, whose ion exchange system (SAC/SBA/MB) is fed by clarified/softened water from the municipal treatment plant. Makeup to the water plant comes from a lake. One December, the city experienced an extremely heavy rainstorm that deposited 7" within 24 hours. This stirred up the lake. Shortly thereafter, large quantities of silica began to appear in the boilers. Investigation revealed that the intense rainstorm stirred up collodial silica, some of which passed through the water treatment plant. Collodial silica does not have an effective charge and is not removed by anion resins. It is also not detectable by the conventional colormetric technique used in many labs. Once the silica reached the boilers, the high temperatures caused a breakdown to reactive silica, which then showed up in lab analyses. RO and UF are strong candidates for makeup treatment at facilities that have to deal with collodial silica. In fact, I heard of one plant that actually retrofitted RO ahead of a demineralizer to minimize this problem.

Practical Example 3-5

This example is taken from the facility whose pretreatment schematic is outlined in Figure 3–12. The RO is equipped with 5-micron filters to screen the iron and manganese oxide particles that are not removed by the pretreatment system. Even with these devices, the membranes showed a large buildup of iron oxides after only 16 months of operation. In fact, the combination of iron oxide and microbiological fouling required a membrane replacement at that time. The two factors caused some membranes to swell so much that they were almost impossible to remove. In one instance, we had to push on the membranes with a 4" by 4" plank and a cherry picker as the power source! The RO was mounted in a truck trailer with only one end available for access. Had we been able to push from the other end, the project might have been easier. This shows the importance of providing access to both ends of an RO.

Practical Example 3-6

Operators of the RO unit above noticed that the 5-micron pre-filters of the RO would typically develop a slimy layer after only two or three days of operation. An analysis revealed microbial counts in the millions per milliliter of sample. As Figure 3-12 illustrates, pretreatment includes chlorine feed eventually followed by dehalogenation with sodium bisulfite. Even though the operators maintain a constant chlorine residual in the makeup water, some microbes survive and then begin to multiply once the chlorine is removed. One possible solution is to feed a non-oxidizing biocide downstream of the bisulfite injection point. Another is to move the sulfite injection point closer to the RO to cut down on the microbe residence time.

Other Membrane Technologies

A number of other membrane technologies are available, some for pretreatment and some for high-purity applications. We will examine the most popular below. First is a discussion of micro- and ultrafiltration, which are most closely related to reverse osmosis. Then, we will take a look at the combination membrane/electrical processes.

Micro- and Ultrafiltration

One of the major membrane system manufacturers (Osmonics) has provided an interesting example of the flow path diameters in reverse osmosis, nanofiltration, and ultrafiltration membranes. RO pores are comparable to a dime on the Pacific Ocean, NF to a quarter, and UF to a half-dollar. Pores in MF membranes may be from one to twenty times smaller than UF pores. These are very small indeed. Even so, RO membranes are the only one whose pore sizes are small enough to remove ions. UF and MF function in a conventional manner and screen particles mechanically. Nonetheless, these techniques have strong practical merit, especially for pretreatment of RO units and other systems. Refer again to Figure 3–28. MF will remove particles such as bacteria and very small dust particles. UF will go even further and remove collodial silica and viruses. In fact, MF and UF are finding applications in the potable water treatment industry as a method for removing harmful organisms from drinking water. One advantage of these systems is that pressure requirements are generally much lower than RO due to the larger pore size.

Membranes may be of the spiral-wound or hollow-fiber type. A number of applications for both types are becoming popular. MF hollow-fiber membranes are now being used in place of clarifiers for suspended solids removal. Combination membrane treatments like UF/Double-Pass RO can supply high-purity water suitable for some boilers. MF and UF are popular in the beverage industry for solids removal from solutions.

Electrodialysis and Electrodialysis Reversal

Electrodialysis (ED) and electrodialysis reversal (EDR) are based upon membrane technology and ionic attraction to an electric field. A simple electrodialysis schematic with the power off is shown in Figure 3–35. Water passes through cells that are separated by either cationic or anionic membranes. When a potential is applied across the membranes, cations are attracted to the cathode, and anions to the anode. Ionic flow is controlled by the membranes. Cations pass through the anion-impermeable membranes and anions pass through the cation-impermeable membranes. This generates either purified or concentrated water within the compartments (Figure 3–36). In the figure, compartments #2 and #4 contain purified water and compartments #1, #3, and #5 contain concentrate. The purified and concentrated streams are collected at the discharge of the compartment, and sent to process and waste, respectively.

CHAPTER 3: MAKEUP WATER TREATMENT

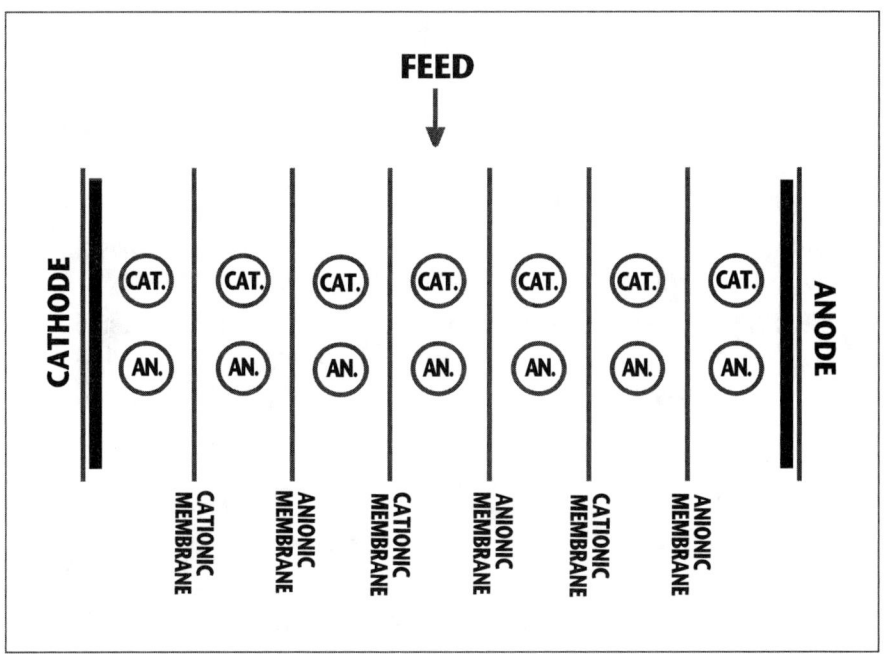

Fig. 3–35 *EDR with Power Off*

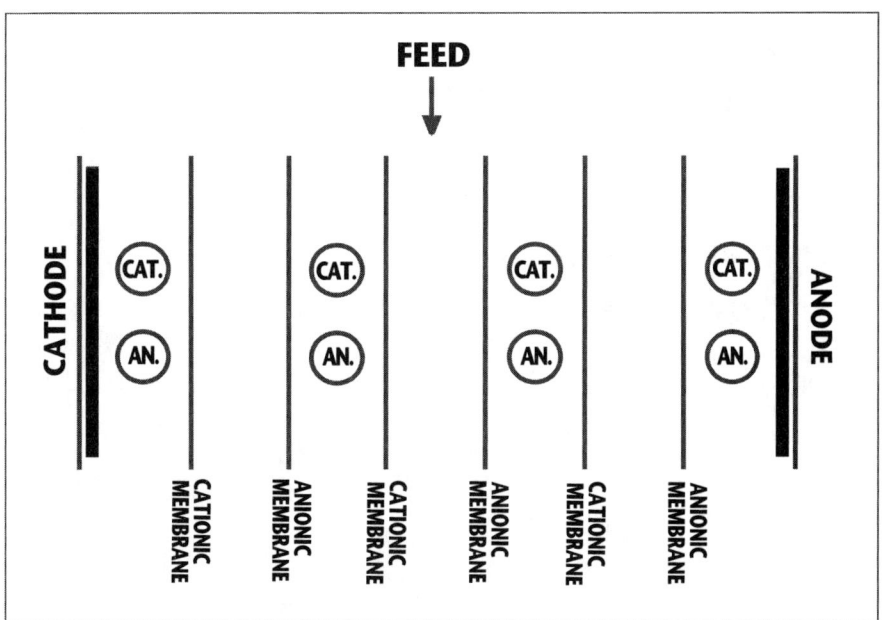

Fig. 3–36 *EDR with Power On*

Flexibility in system design is possible by varying the number of compartments. Many compartments placed together comprise a stack. The number of compartments per stack and number of stacks can be selected to produce the required quantity of water.

The electrodialysis process has been modified into electrodialysis reversal, in which the polarity of the field is regularly alternated. This reduces the formation of deposits on the membranes.

EDR systems are capable of removing up to 85 percent of influent dissolved solids with recoveries of 75 to 85 percent. The process is not good at removing silica, as silica is only very slightly ionized and is not strongly attracted to the electrodes. EDR offers several advantages, most notably:

- Water is not forced through the membranes as it is in RO, so the membranes are less susceptible to fouling.

- EDR membranes are usually constructed of durable material such as polysulfone. This material is resistant to oxidizing biocides, and in fact, maintaining an oxidant residual in the EDR helps keep the membranes free of deposits.

The cost of an EDR system is roughly 30 to 50 percent higher than that of a reverse osmosis unit. EDR is not extensively used for boiler makeup at this time, being more popular in the potable water industry. EDR has also shown promise as an ion concentrator for zero discharge systems.

Electrodeionization

The electrodialysis process has been carried a step further with the development of electrodeionization (EDI). Essentially, these are ED systems to which mixed cation and anion exchange resins have been added (Figure 3-37). EDI systems offer several advantages, but two stand out:

- The addition of ion exchange resin greatly improves the process with regard to silica removal.

- The electric field generates H^+ and OH^-, which regenerate the resin on-line. This eliminates acid and caustic regeneration.

Chapter 3: Makeup Water Treatment

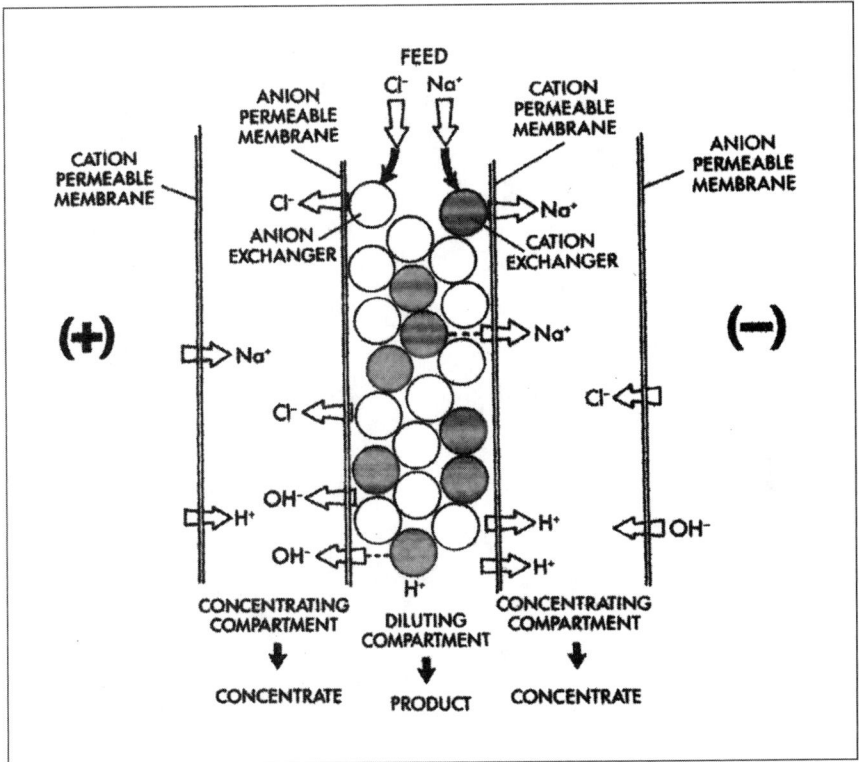

Fig. 3-37 *The EDI Process*
Courtesy of U.S. Filter

An EDI system can produce water of a mixed-bed quality. The units are primarily used as polishers, and a system composed of a reverse osmosis unit followed by an EDI unit might be ideal for generating high-purity water with no chemical regeneration. EDI systems must be protected against the intrusion of suspended solids, because the resin cannot be backwashed as is possible in conventional ion exchange units.

Conclusion

Production of makeup water suitable for feed to steam-generating units requires many different processes. A number of options are available to produce high-purity water, and some treatment method is always possible. Proper design and operation are essential for good operation.

Learning Aide 3-1

ALTERNATE DISINFECTION METHODS

Although chlorine or other oxidizing biocides are the most common disinfectants, other methods are also available for sterilizing water. One of these is ultraviolet light. UV light at a wavelength of 254 nanometers can be a strong disinfectant, especially when used in combination with other techniques. Figure 3–38 shows a pretreatment system at an electronic parts manufacturer. This system is quite a bit more complex than others in this book, because the water must be extremely pure so as not to contaminate any of the very sensitive electronics as they are being manufactured. Let's briefly examine the stages in the treatment process. Following media filtration and preheating is the first of four UV treatments. In the event of a significant microbiological problem, plant personnel can inject sulfuric acid, iodine and a non-oxidizing biocide for supplemental protection. Next comes reverse osmosis as a first-stage purification step followed by two more UV light treatments. Following is fine filtration with a 0.3-micron filter unit and then demineralization to reduce dissolved ion concentrations to ppb levels. Last is yet another UV treatment.

Some of this equipment may appear to be redundant. After all, reverse osmosis removes microbiological organisms. But, the parts that are being manufactured are so sensitive that precautions must be taken to prevent any fouling. Thus, the additional UV treatment systems.

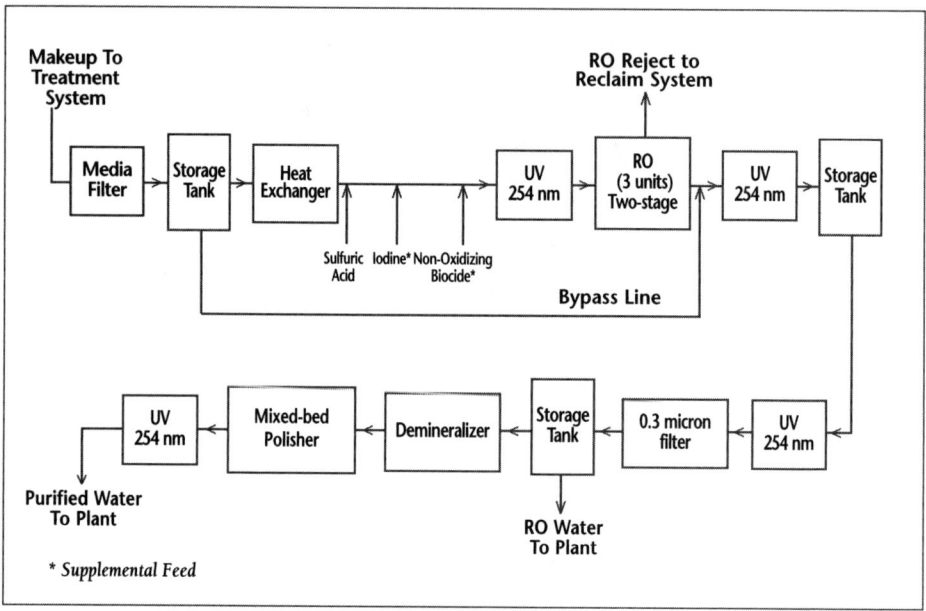

Fig. 3–38 *Flow Diagram of an Actual Makeup Water System Showing UV Treatment Locations*

Learning Aide 3-2

EXAMPLE OF FILTER MEDIA SELECTION

Listed below are the actual specifications for filter media in a utility pre-treatment system. I include this example to show what multimedia filtration can actually be. The list shows the bed from top to bottom. Note that the lightest, coarsest material is on top.

Pre-treament Filter Media Specifications

Media	Particle Size	Depth
Anthracite	0.85 to 0.95 mm	18 inches
Quartz Sand	No. 20	8 inches
Garnet	30 to 40 mesh	4 inches
Quartz Support	Bed Graded *	24 inches

* Support bed contains three layers of quartz gravel, 0.5" x 0.75", 0.25" x 0.5", and 0.125" x 0.25".

Table 3–6 *Pre-treatment Filter Media Specifications*

Learning Aide 3-3

Sulfuric Acid and Caustic Specifications

Sulfuric acid and sodium hydroxide used for regeneration must be of a reasonable purity for proper performance. Should any readers ever be tasked with acquiring these materials, the following specifications can be used as a guide.

Sulfuric Acid (Contaminant Maximum Concentrations)
- Iron – 20 ppm
- Copper – 0.5 ppm
- Manganese – 0.2 ppm
- Chloride – 10 ppm

Sodium Hydroxide (Contaminant Maximum Concentrations)
- Iron – 10 ppm
- Silica – 10 ppm
- Chlorates – 100 ppm
- Sodium Carbonate – 0.5%
- Sodium Chloride – 0.5%
- Sodium Sulfate – 0.5%

Learning Aide 3-4

SILT DENSITY INDEX

The fouling index of water supplies may be measured by several different techniques. One is a direct analysis of the suspended solids concentration. Another is turbidity, which we learned is a common measurement for clarifier and filter effluents. An analysis developed for reverse osmosis systems is the silt density index (SDI). The test for SDI is relatively simple; water is filtered through a 0.45 filter at 30 psig. Measurement is taken of the time for 500 ml of water to pass through the filter at the beginning of the test and again after 15 minutes. The SDI is calculated by the following equation:

$$SDI = \frac{1-(t_i/t_f)}{T} \times 100 \qquad [Eq. 3\text{--}13]$$

where,

t_i = Time for initial 500 ml sample to pass through filter

t_f = Time for final 500 ml sample to pass through filter

T = Time between measurements (15 minutes is standard)

As an example, consider the following actual readings taken from the influent of an operating reverse osmosis unit.

t_i = 34 seconds

t_f = 66 seconds

T = 15 minutes

The SDI calculates to 3.2.

As a general guideline, the SDI should be at least below five and ideally three or below. However, SDI should not be the only criteria that determines the suitability of an RO application or the efficiency of a pretreatment system. The type of water or nature of contaminants must also be considered. Some reverse osmosis units have been known to operate well with SDIs around or above the 5.0 reading, while others have exhibited problems treating raw waters with much lower SDIs. SDI readings of the RO outlined in Practical Example 3-5 almost always ranged between 1 and 3, yet the membranes fouled with iron oxide particles.

The major water treatment equipment manufacturers will perform SDI tests, and some sell SDI test kits. The greatest problem I discovered with the test kits is making sure that the filter is loaded in properly so that the sample is spread evenly over the entire surface area of the filter. Experience has also shown me that one must take multiple SDI readings, ideally under different influent conditions, to accurately determine the SDI or range of SDI of the water.

FUNDAMENTALS OF STEAM GENERATION AND SOURCES OF CONTAMINATION 4

A basic law of physics says that energy is neither created nor destroyed, only transformed. I will not try to analyze this from the nuclear physics angle, but for everyday life this statement is certainly true. Common examples are all around. When an automobile stops, the brakes grow warm as mechanical energy changes into thermal energy. Wind is generated by the sun's energy, which is a transformation of thermal energy into mechanical energy. Even in simple combustion turbines, multiple energy transformations take place. Burning fuel (thermal energy) heats gases, which expand and flow (mechanical energy) through turbine blades to generate electricity.

The energy transformation at a steam-generating power plant is similar, with the obvious exception that steam rather than combustion gases drive the turbine. All of the subsystems that comprise a steam-generating unit are designed to maximize energy efficiency. We will look at those systems or equipment directly involved in the production or transport of steam and water. It is important for the student to have an understanding of the fundamental steam-generating network and basic water/steam chemistry details. This chapter hopefully provides that background.

With this knowledge the reader will be able to more clearly understand the next two chapters, which discuss treatment programs and monitoring.

An Overview of Problems Caused by Water or Steam Contamination

Contaminants may enter steam-generating systems from a number of sources, or they may be generated within the system. Impurities cause many problems. Hardness ions can precipitate as carbonates, hydroxides or silicates to form scale in boiler tubes. Oxygen in-leakage through a condenser can corrode downstream condensate and feedwater piping, and boiler tubes. Silica will volatilize in the boiler and carry over to the turbine, where it deposits on turbine blades. Iron corrosion products in general account for the bulk of deposits on boiler tube walls. Many other difficulties are also possible, as we shall see.

Before launching into a discussion of the steam generating system and sources of contamination, I want to review the basic types of corrosion. After all, corrosion is what we chemists are trying hardest to prevent in steam-generating systems—that and scaling/fouling. Much of the discussion in the next section is adapted from information provided over the years by the Nickel Development Institute (NiDI) with supporting information from the National Association of Corrosion Engineers (NACE) International. The NiDI is a nonprofit organization, which promotes the use of nickel as an alloying material, has published many excellent technical documents on power plant materials performance and selection. Many of these documents are free of charge. I recommend that any readers who wish to learn more about power plant materials science contact the NiDI or NACE International for more information:

Nickel Development Institute
214 King Street West, Suite 510
Toronto, Ontario, Canada
M5H 3S6
Voice (416) 591-7999
Fax   (416) 591-7987

NACE International
1440 South Creek Drive
Houston, TX 77084
(281) 228-6200

CHAPTER 4: FUNDAMENTALS OF STEAM GENERATION
AND SOURCES OF CONTAMINATION

Corrosion Fundamentals

Many types of corrosion are known. Some, however, are much more common than others. Eleven of the most important corrosion mechanisms concerning steam generation are these:

- General corrosion
- Pitting
- Crevice corrosion
- Galvanic corrosion
- Erosion-corrosion
- Stress corrosion cracking
- Corrosion fatigue
- Intergranular corrosion
- Dealloying
- Exfoliation
- Microbiologically-influenced-corrosion

Each of these corrosion mechanisms can cause serious problems, and we will see examples of many of them throughout the remainder of the book. Learning Aid 4–1 lists a number of the most common alloys (and their compositions) that are used as materials of construction for steam-generation plants. As we progress through the remaining chapters, we will see how these materials perform in corrosive environments.

General Corrosion

General corrosion occurs when a metal is attacked uniformly by a corroding substance. Rust is one such example. The corrosion of active metals by acid is another. Pipes, tubes, and other power plant materials are always designed with a general corrosion safety factor. Since general corrosion is spread out over a wide area, metal integrity may remain solid for many years. Unfortunately, many corrosion mechanisms are not general in nature but are localized. These are much more serious.

Pitting

Pitting is a localized form of attack in which the corrosion penetrates downwards through the metal. Pitting is a most insidious mechanism and may cause failures even when the bulk of the material is still solid. Several types of pitting are possible. One of the most common is under-deposit corrosion, which is a phenomenon that occurs in boilers and cooling water systems. Deposits in cooling water systems generate oxygen differential cells, in which the environment underneath the deposit becomes oxygen deficient in comparison to the bulk solution. This causes the tube metal below the deposit to become anodic to clean surfaces, where oxygen is in good supply. Under-deposit corrosion is one of the primary reasons why it is so important to prevent fouling and scaling in water systems. Deposits in boilers may produce oxygen differential cells, or they may trap corrodents underneath that attack the boiler tubes directly.

Another form of pitting is an attack common to stainless steels. As Learning Aide 4–1 points out, stainless steels develop a protective oxide layer that protects the base metal. In certain steels this oxide layer may be penetrated by impurities in the water, which induce pitting. Chloride is a particularly notorious pitting agent, and it will attack some grades of stainless steel, including one of the most commonly used heat exchanger materials, 304 stainless. As the attack continues, the chloride concentration in the pit increases and the pH goes down, further exacerbating the problem.

Permit me to digress here for a minute about stainless steels. They are widely used throughout the power, chemical, and manufacturing industries for piping, vessels, heat exchangers, and other equipment. They serve a valuable function. However, if they are selected improperly or for the wrong application, corrosion failures can occur very rapidly. Disruption of the protective oxide layer is a common problem, and corrosion can become severe if the material is used in a naturally aggressive environment. I have seen case histories where construction workers, during material installation, scratched the surface of the metal with a screwdriver or similar object. The imbedded iron and disruption of the protective oxide layer set up a localized corrosion cell. I have seen other examples where workers wrote on stainless steel panels using a grease pencil. The panels were then installed in a corrosive environment. The steel underneath the writing then became anodic to the remaining metal and corroded.

Crevice Corrosion

Crevice corrosion can almost be thought of as pitting, but in this case a mechanical factor, the physical joining together of two metals, leads to sites (crevices) that become oxygen depleted. The crevices become anodic to the exposed metal surface and corrode in a similar mechanism to the under-deposit process.

Galvanic Corrosion

Galvanic corrosion occurs when two dissimilar metals are coupled together in a corrosive medium. The more passive of the two serves as a cathode and the more reactive becomes an anode. Galvanic corrosion can be especially severe when a small anode is coupled with a large cathode. One of the classic examples is of copper plates connected by steel rivets in seawater service. Corrosion of the rivets is intense. The reverse situation, copper rivets holding steel plates, is much better from a corrosion standpoint. Here we find a large anode coupled to a small cathode. Although corrosion of the steel still occurs, metal loss is much less localized and severe.

Erosion-Corrosion

Erosion-corrosion is a combination of physical and chemical attack. It occurs where a flowing stream erodes the outer layer of the metal exposing fresh metal to further corrosion. High fluid velocity or disturbances in flow patterns may initiate erosion-corrosion. A special form of erosion-corrosion, flow-assisted-corrosion, is causing great concern in the power industry. Heat recovery steam generators may be particularly susceptible to this attack.

Stress Corrosion Cracking

Stress corrosion cracking is another mechanical/chemical mechanism. The corrosion occurs at stress points in metals, which are attacked by a specific corrodent. The 304 and 316 stainless steels offer a good example. When placed under stress in a corrosive environment, such as a solution containing chlorides, the material will begin to corrode at the stress points. Turbine blades are prime locations for SCC, as they are constantly under stress.

Corrosion Fatigue

Have you ever broken a wire by bending it back and forth repeatedly until it failed? This is an example of fatigue. Failure was caused by cyclical stress. Cyclical stresses are common in rotating machinery. Couple the stresses with an aggressive environment and corrosion may occur.

Intergranular Corrosion

Intergranular corrosion occurs along the grain boundaries of metals. This attack is often set up by poor heat treatment of the metal. Improper heating or cooling of an alloy may occur during the fabrication process, or more often, when the material is being welded in the field. This can alter the crystalline structure of the alloy and reduce corrosion resistance. Many stainless steel failures have occurred due to poor welding, wherein the heat from the process caused chromium to combine with carbon into small globules. This chromium carbide precipitation reduces chromium's ability to form a protective oxide layer on the steel surface and opens the way for corrosion.

Dealloying

Dealloying is the selective leaching of one metal from an alloy. Two of the most common forms of dealloying are dezincification and denickelification of Admiralty metal and copper-nickel alloys, respectively. Although some debate exists as to the exact corrosion mechanism, the results are similar. The zinc or nickel depart leaving a spongy copper mass behind.

Exfoliation

Exfoliation is a readily observable phenomenon. Flakes of metal appear on the pipe or tube surface. These flakes will often wash downstream and foul other areas of the steam generating system. Exfoliation may be the result of mechanical factors more than chemical attack.

Microbiologically-Influenced Corrosion

Microbiolgically-influenced-corrosion, or MIC for short, is a common problem in cooling water systems. We will examine this phenomenon more closely in chapter 7. Microbes not only form deposits that may cause under-deposit corrosion, but many organisms also secrete corrosive chemicals as part of their metabolic processes. These chemicals will directly attack the base metal.

The Steam Generating Network

Now that we have briefly reviewed the fundamentals of corrosion, let's look at the steam-generating network and examine the chemistry and the problems which can occur due to contamination or poor chemical treatment.

Figure 4–1 illustrates a steam-generating network that might exist at a co-generation facility. The principal components or processes include a boiler, steam turbine for generating electric power, steam feed to and condensate return from an industrial process or processes, a condenser for recovering turbine exhaust steam, a condensate/feedwater system for returning collected condensate to the boiler, a polisher to purify the returned condensate, and a water treatment system to provide purified makeup to the boiler.

Makeup Water

Chapter 3 outlined the fundamentals of the most common makeup water treatment systems, so we will not examine these operations in much more detail. As we have seen, upsets in the makeup water treatment unit can introduce a variety of impurities into the condensate. With ion exchange systems, the principal contaminants are either sodium or silica, unless a gross overrun occurs. Silica leakage is the most troublesome, as silica will carry over directly from the boiler into the turbine and form deposits on the blades. A well-designed monitoring system can prevent most makeup treatment chemistry upsets.

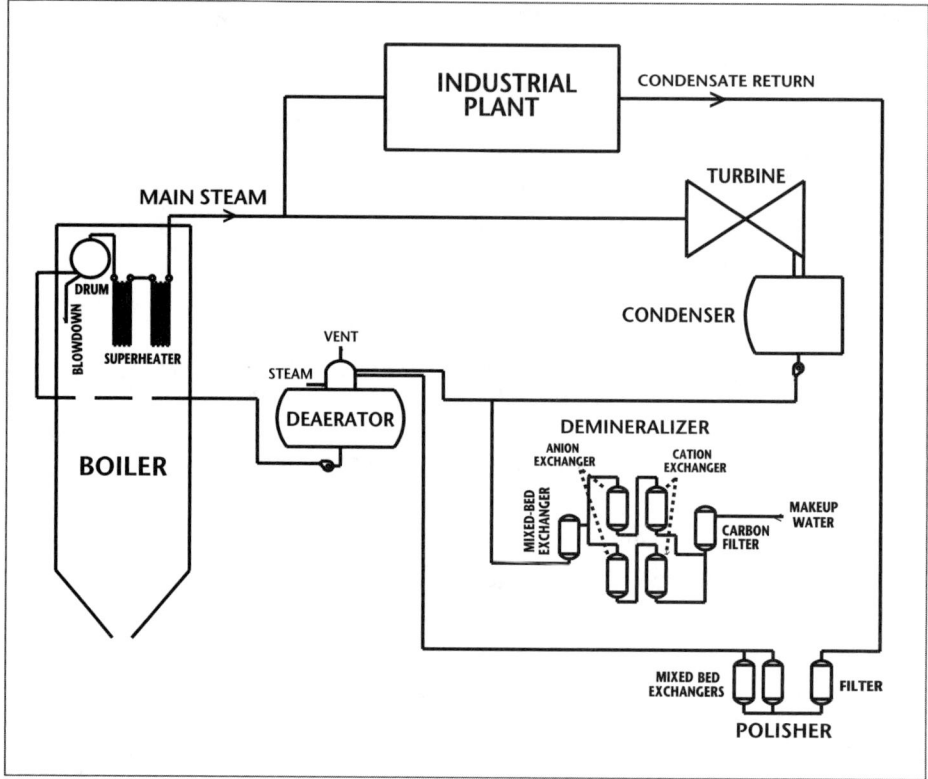

Fig. 4–1 Possible *Steam Generating Network at a Co-generating Facility*

Condensate Return from Industrial Processes

Steam is used as a heating agent for many industrial processes. It is usually economical to return the condensate to the boilers to save on makeup water costs. However, the condensate may pick up a variety of impurities on its passage through heat exchangers and piping. Potential contaminants include these:

- Inorganic ions
- Organic compounds including oils
- Particulates caused by corrosion in the condensate return lines

CHAPTER 4: FUNDAMENTALS OF STEAM GENERATION AND SOURCES OF CONTAMINATION

Each of these will cause difficulties in the boiler. Inorganic ions may react with other impurities to form scale. Organic contaminants can cause foaming in a boiler, char onto the surface of waterwall tubes, and break down into organic acids that carry over to the turbine and cause corrosion. Practical Example 4–1 illustrates a very troublesome case of organic contamination. Particulates may foul downstream equipment and form deposits in boiler tubes. In fact, iron oxide deposition is the single greatest contributor to deposit buildups in high-pressure boilers.

A particularly troublesome form of corrosion in condensate return lines is carbonic acid attack of mild steel pipe. Consider for a moment a low-pressure boiler whose makeup treatment only consists of sodium softening. Other constituents in the water, and in particular anions including bicarbonates, pass directly to the boiler. Aqueous bicarbonate solutions, when heated, react as follows:

$$HCO_3^- + heat \rightarrow CO_2\uparrow + OH^- \qquad [Eq.\ 4\text{–}1]$$

The carbon dioxide carries over with steam. When the steam condenses, some of the CO_2 combines with water to produce carbonic acid:

$$CO_2 \rightleftharpoons H_2O + H_2CO_3 \qquad [Eq.\ 4\text{–}2]$$

This lowers the pH and induces general corrosion. However, carbonic acid will also directly attack carbon steel, which corrodes the metal via the following mechanisms:

$$Fe + 2H_2CO_3 \rightarrow Fe(HCO_3)_2 + H_2 \qquad [Eq.\ 4\text{–}3]$$

$$Fe(HCO_3)_2 \rightarrow FeO + 2CO_2\uparrow + H_2O \qquad [Eq.\ 4\text{–}4a]$$

$$4Fe(HCO_3)_2 + O_2 \rightarrow 2Fe_2O_3 + 8CO_2\uparrow + 4H_2O \qquad [Eq.\ 4\text{–}4b]$$

As these equations illustrate, iron is dissolved by carbonic acid to form an iron bicarbonate complex, which in turn breaks down or reacts with oxygen to release CO_2 and start the process over again.

Carbonic acid corrosion often causes channeling of the pipe wall immersed in the condensate. The corrosion will penetrate the wall over time. Oxygen will also cause corrosion in condensate return lines, but it is more irregular and appears as pits. Good deaeration practices in the steam-generating system help minimize the transfer of oxygen to the condensate.

Practical Example 4-1

This example is taken from an organic chemical manufacturing plant, where the bulk of process steam is produced by four, 100,000 lb/hr, 550 psig package boilers. The boilers are equipped with superheaters, and the steam is used for production of phenol (C_6H_7O) and phenol-derivative compounds. Approximately 80% of the condensate is returned to the boilers. Neutralizing amines are injected into the condensate for corrosion control. Steam is only used for process functions and does not drive any turbines. The laboratory is very rudimentary, and operators only conduct a few simple grab-sample analyses. The only organic compound they can analyze is phenol, as they do not have the equipment or procedures to test for any of the other organics that may come back with the condensate.

Sporadic, but abnormally frequent tube failures have been a regular occurrence in the superheaters of the four boilers. When a tube fails, the company removes and replaces the entire superheater, a rather expensive procedure. Tubes taken from discarded superheaters exhibit internal deposits that range from perhaps $1/8$ to $1/4$ inches in depth. An analysis of these deposits showed that they were primarily composed of iron oxide, silica, sodium, and carbon. All of the boiler drums contain steam separating internals, and the company added an external set of separating devices to one of the most problematic boilers. The problems, however, still persist.

Water chemistry records revealed the probable source of the difficulties. For boilers of this pressure, the recommended maximum TOC concentration in the boiler water is 0.5 ppm. Prior water chemistry records by an outside vendor indicated that TOC levels in the condensate return sometimes reached 20 ppm. Plant personnel admitted that other spot tests had shown TOC concentrations as high as 200 ppm in the condensate return! This data immediately suggested that foaming was probably occurring in the drums, and indeed, when a co-worker

Chapter 4: Fundamentals of Steam Generation and Sources of Contamination

and I examined the drum water sample line discharge from one unit, foam was clearly evident.

Several corrective actions immediately suggest themselves. These include the following:

- Upgrade the sampling system so that accurate steam samples are available. The current sampling taps have been placed in locations that do not provide representative samples.

- Upgrade analytical procedures so that plant personnel or an outside vendor can determine all of the major constituents in the condensate return, boiler feedwater, boiler water, and steam. For example, of the five major organic chemicals produced at the facility, the operators can only analyze for one. The procedure involves a colorometric determination, and the operators reported that the sample often develops a different color than that outlined in the analysis procedure. This indicates the presence of other compounds.

- Consider installation of a condensate polishing system. System design would be based on the constituency and concentration of contaminants that are being introduced to the boilers. Since in all likelihood the major contamination is due to phenols and phenol-derivatives, plant managers should talk with reputable polisher manufacturers to discuss applicable treatment methods. Disposal of any used resin or absorbant is another important consideration..

Steam Surface Condenser

The steam surface condenser is an integral component at most electric utilities and at large industries that generate power (A few use air-cooled condensers). The condensation process greatly improves turbine efficiency (see Learning Aide 4–2). The operation of a condenser is not extremely complicated. Exhaust steam from the turbine passes over water-cooled tubes, which cause the steam to condense. The condensate collects at the bottom of the condenser in the hotwell, from which it is withdrawn for return to the boiler. Gases that enter with the steam are pulled out of the condenser through one or more air-removal compartments. Makeup water is usually introduced to the unit through the condenser. The condenser may also act as a collection point for the drips from low-pressure feedwater heaters.

The basic components of a condenser are the containment shell, cooling water tubes, inlet and outlet cooling water boxes, the hotwell, and an air-removal system. Any reasonably sized condenser will contain several thousand tubes. These are rolled or otherwise permanently attached to a tube sheet at the inlet water box. The tubes are supported within the condenser by tube support plates located at regular intervals. Condensers may be single- or double-pass and split case. In a single-pass condenser, water flows through the tubes, collects at the outlet water box, and then discharges to a receiving source. In the very common double-pass condenser, the water reverses course through a U-shaped turnaround, and then flows through another set of tubes. The second pass is located below the first, and both share the same tube support plates. Split case condensers, as their name implies, handle a divided water flow. It is sometimes possible in split-case condensers to isolate one side and search for a leaking tube while the other side remains in operation.

Steam condensation produces a very strong vacuum, which draws air into the condenser through any crack or opening. Accordingly, all condensers are equipped with an air-removal system. The outline of an air-removal compartment is shown in Figure 4–2. This is a drawing of the tube layout of one-half of an actual split-case condenser. The shroud extends from front to back of the condenser, and is open at the bottom. A vacuum is applied to the compartment through a vacuum pump that ejects gases to the atmosphere. The other half of this condenser contains an identical air-removal compartment.

Condenser tubes must exhibit good heat transfer to maximize efficiency. The most common tube materials are 304SS, 90-10 and 70-30 copper-nickel, and Admiralty metal. Newer materials, especially those for seawater cooled condensers, include titanium and the 6-moly stainless steel alloys. Each offers advantages and disadvantages that will present themselves as we examine tube failure mechanisms.

Chapter 4: Fundamentals of Steam Generation and Sources of Contamination

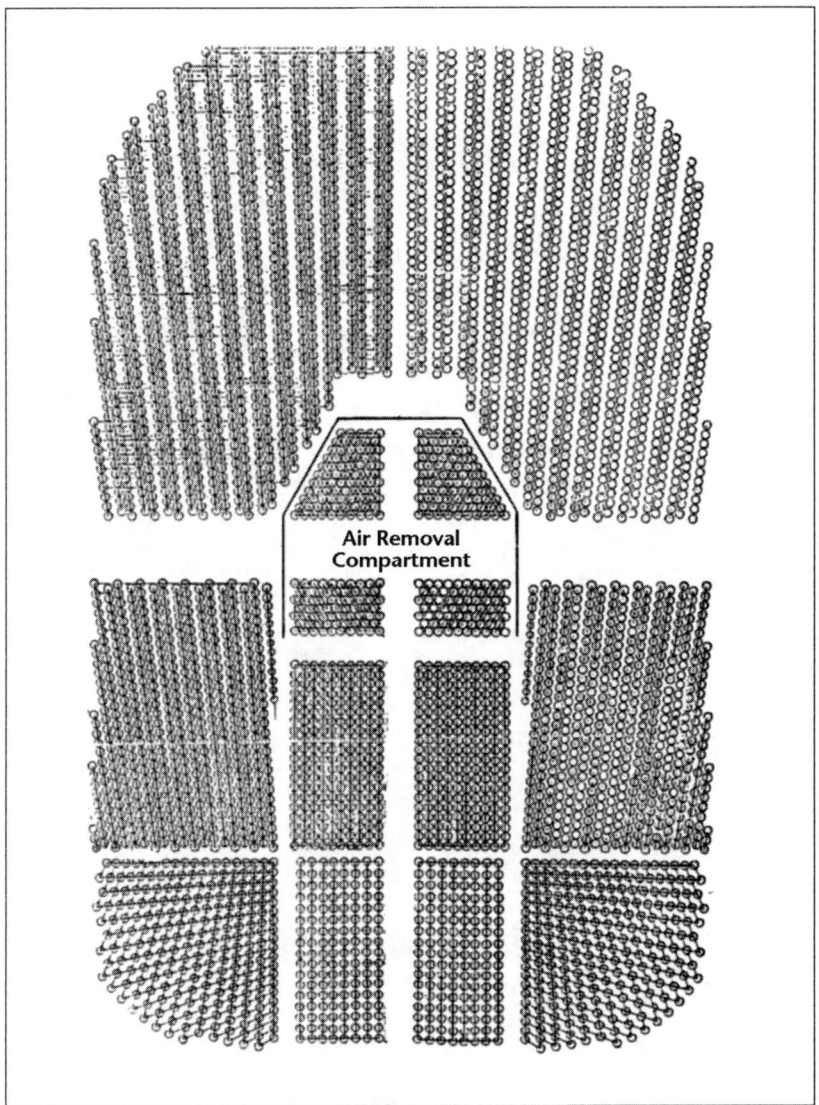

Fig. 4–2   *Condenser Tube Map Showing Air Removal Compartment*

In chapter 2, we looked at the major impurities in natural water supplies. Any cooling water contains much greater concentrations of dissolved (and sometimes suspended) materials than boiler condensate, so even small leaks can cause gross contamination of a steam-generating system. In condensers supplied by open recirculating systems (cooling towers) or seawater, the effects are even worse. Let's review again some of the major dissolved ions in cooling water. These are outlined in Table 4–1. When these impurities enter a steam-generating system, the high temperatures can initiate many reactions. Several of these mechanisms are outlined below.

- Calcium will precipitate with a number of anions, including carbonate and sulfate, to form scale on boiler tubes. Scales generate under-deposit corrosion and restrict heat transfer. Even in the absence of severe corrosion, the tubes may fail due to overheating.

- Reactions of some salts with water may lead to direct corrosion. A classic example is the reaction of magnesium salts with water at high temperatures.

$$MgCl_2 + 2H_2O + heat \rightarrow Mg(OH)_2 + 2HCl \qquad [Eq.\ 4\text{–}4b]$$

  This reaction can lower boiler water pH to less than 5.

- Silica may precipitate by itself on boiler tube walls, or it may combine with magnesium or other ions to form additional troublesome scales. Silica will carry over with steam and precipitate in turbines and reheater lines.

Major Dissolved Ions in Cooling Water

Cations	Anions	Other
Calcium	Bicarbonate	Silica
Magnesium	Chloride	Iron
Potassium	Nitrate	Manganese
Sodium	Sulfate	Organics

Table 4–1 *Major Dissolved Ions*

CHAPTER 4: FUNDAMENTALS OF STEAM GENERATION
AND SOURCES OF CONTAMINATION

The following two practical examples show actual histories of the great difficulties caused by cooling water in-leakage to a condenser. I wish to give you readers a word of warning about this section on condensers. For five years during my career at an electric utility, I was responsible for condenser performance monitoring. You will see a lot of examples in this section and in chapter 6, because I have a lot of information to pass along. I can tell stories that will curl your hair. Okay, maybe not that drastic, but to plant managers such events can be frightening.

Practical Example 4-2

An 80 MW unit supplied by a 1250 psig coal-fired, cyclone boiler had just been returned to service from a scheduled autumn outage. Laboratory personnel discovered that a condenser leak was allowing contaminants to enter the system, such that condensate total-dissolved-solids (TDS) concentrations at times reached 0.75 ppm. Although the lab staff requested that the boiler be taken off line immediately, the operations managers refused due to load demand issues. The boiler was on congruent phosphate control, so the lab staff increased monitoring frequency and attempted to maintain the boiler water treatment program within recommended guidelines. After approximately three weeks, operators discovered the source of the leak and corrected the problem.

Two months later, boiler waterwall tubes began to fail with alarming frequency. The unit came off numerous times for tube repairs, and in at least one instance had only been back on-line for a few hours when another tube failed. The failures happened so regularly that plant management scheduled an emergency tube replacement during an upcoming spring outage. The repairs cost over $2,000,000. The mechanism attributed to these failures was under-deposit corrosion caused by excessive sludge and scale formation. Interestingly, the leak was not from a failed condenser tube. The condenser hotwell is equipped with a drain line that discharges to the cooling water outlet tunnel. During the autumn outage, an operator opened the line to drain the hotwell but then forgot to close the isolation valve before startup. Once the unit went on-line, the strong condenser vacuum pulled cooling water into the hotwell.

Practical Example 4-3

This problem occurred in a 200 MW unit with a 2400 psig, coal-fired tangential boiler. At the time of this event, the only on-line chemistry monitoring for the unit consisted of a sodium analyzer at the condensate pump discharge (CPD). The monitor was not equipped with an alarm. On the morning of the upset, an operator checked the monitor at 7:00 a.m., and found the sodium concentration to be less than one part-per-billion (ppb). When lab chemists came by at 7:45 a.m., the monitor needle was pegged beyond the upper limit of 100 ppb. Lab personnel immediately grabbed a boiler water sample and found to their horror that no treatment chemicals remained in the sample and that the pH had dropped to 5.8! The chemists immediately notified the operations managers, who took the unit off as quickly as possible. During this time the chemists injected treatment chemicals into the boiler water to raise the pH and stabilize the chemistry. The operators opened the blowdowns on the lower headers to remove solids.

Once the unit came off-line, maintenance personnel discovered the problem. A plug had fallen out of a topmost condenser tube, which had previously failed due to steam erosion. The tube had massive failures throughout its length, so large quantities of cooling water poured into the condenser. Even though this condenser is on a once-through cooling system supplied with relatively soft surface water, the massive size of the leak quickly consumed the boiler water treatment chemicals. This story had a good ending. The quick action by the plant staff prevented serious scale formation or corrosion. The boiler was chemically cleaned at the earliest available opportunity. No tube failures occurred after the upset. However, the event proved to be a catalyst in convincing utility managers to install a comprehensive on-line water chemistry monitoring system.

The potential sources of contaminant ingress within the condenser are immense. Tube leaks develop for a variety of reasons. The most common include these:

- Waterside corrosion.
- Steam-side corrosion.
- Steam impingement of upper tubes.

- Failure where the tubes are rolled or mechanically attached to the tube sheet.
- Failure of tubes at tube support plates due to vibrational effects from flowing steam.

Let's examine these mechanisms.

Waterside Corrosion

Waterside corrosion is often caused by poor microbiological treatment of the cooling water. Once microbes attach to tube walls they secrete a sticky protective film that increases the deposit volume. This layer in turn absorbs silt and other debris from the cooling water, further increasing the deposit mass. Several corrosion mechanisms are possible. Deposits in general may cause corrosion due to oxygen differential cells that develop between the aerated cooling water and the oxygen-deficient zone beneath the deposit. Also, many microorganisms produce acids as part of their metabolic processes, and the acids, which are trapped beneath the deposit, directly attack the base metal. This is known as microbiologically-influenced-corrosion (MIC), an example of which is shown in Practical Example 4–4.

Scale deposition can also cause under-deposit corrosion due to the formation of differential oxygen cells. Sometimes the tubes may be directly attacked by contaminants in polluted waters. Ammonia combined with oxygen is corrosive to copper. Although the ammonia concentration of many natural water supplies is low, manmade influences can come into play. A supply located downstream from the discharge of a wastewater treatment plant may contain higher quantities of ammonia. Agricultural runoff during rainy periods may raise the ammonia levels of surface supplies. Sulfide is a severe copper corrodent. An interesting example of sulfide corrosion is outlined in Practical Example 4–5.

Practical Example 4-4

The following occurred in a condenser tubed with 304 stainless steel. The unit was down for a scheduled maintenance outage. The operators were not told to drain the waterside of the condenser during the outage, so many of the tubes remained immersed in standing water for about a month. When the unit came back on line, cooling water in-leakage was severe. Upon inspection, plant personnel discovered numerous pinhole failures in the tubes. The suspected failure mechanism was MIC, induced by the stagnant conditions. An outside contractor was brought in to remove and replace all 15,000 tubes in the condenser. In a somewhat natural human reaction, plant management did not retube with stainless steel, but selected copper-nickel (90-10 in the main section and 70-30 in the air-removal section) as the replacement. Stainless steel could have been used again as long as proper layup procedures were followed. Copper-alloy tubes can also suffer from MIC, but copper-based materials are usually more resistant to microfouling because copper ions are toxic to aquatic organisms. A once-common treatment to control algae growth on ponds consisted of spreading copper sulfate on the pond surface.

Practical Example 4-5

This condenser was originally equipped with Admiralty tubes, which had, after 20 years, begun to fail due to steam-side ammonia/oxygen attack. (See Practical Example 4–6.) Plant personnel decided to replace the tubes in the affected zones with 90-10 copper-nickel. This material had performed well in other condensers. Within two years, the new tubes began to fail. When a tube was pulled for examination, numerous pits were observable on the waterside of the tube. One half-section sample, cut to one foot in length, contained eight, through-wall penetrations. The corrosion was traced to the manufacturing process. The supplier used a lubricant that contained sulfide, which is a very corrosive agent to copper. The supplier did not clean the tubes before they were installed. The sulfide kept eating through the tubes until they failed.

Steam-Side Corrosion

Steam-side corrosion can also be insidious. The most common form of steam-side corrosion results from ammonia-oxygen attack of copper alloys, particularly Admiralty metal. Ammonia is generated in a boiler system through the decomposition of nitrogen-based oxygen scavengers or neutralizing amines added for feedwater pH control. Or, ammonia itself may be directly injected for pH control. Much of it carries over with steam. The following practical example describes a classic case of ammonia attack.

Practical Example 4-6

This condenser was tubed with Admiralty metal, and had operated for 17 years with few tube failures. Without warning, periodic failures began to occur in the air-removal section. These failures caused a number of forced outages. During one of the outages, maintenance personnel pulled out four tubes that had previously been plugged. Each of the four tubes exhibited circumferential gouges at interface points between the tubes and tube support plates. The gouges showed many small cracks, and one or more of the gouges in each tube contained a through-wall penetration. A metallurgical firm identified the corrosion as ammonia-oxygen attack of the Admiralty metal. Research by the NiDI indicated that 70-30 copper-nickel alloy was much more resistant than Admiralty to ammonia attack, so utility management had the air-removal section retubed with this material. This type of corrosion usually occurs within or directly below the air-removal section where ammonia and oxygen concentrations are highest. The general corrosion mechanism proceeds as follows:

- In the absence of excess oxygen, the walls of copper-alloy tubes form a layer of cuprous oxide, Cu_2O. This layer is reasonably protective of the underlying base metal. Copper atoms in cuprous oxide exist in a +1 oxidation state.

- Excess oxygen further oxidizes the layer to cupric oxide, CuO. Copper atoms now exist in a +2 oxidation state.

- Ammonia is a strong complexing agent towards Cu^{+2}, and it reacts with the ion as follows:

$$Cu^{+2} + 4NH_3 \rightarrow Cu(NH_3)_4^{+2} \qquad [Eq.\ 4\text{--}6]$$

- The ammoniated copper complex washes away.

Ammonia is very soluble in water. As steam condenses in the air removal section, ammonia dissolves in the condensate. The condensate tends to collect and run down the tube support plates, so ammonia concentrations are very high at the interface between the tubes and support plates. Ammonia-oxygen attack of copper-alloy condenser tubes has been frequently reported in the literature. Remember the discussion of coordination complexes in chapter 1? Here is the example I promised to mention.

Steam Impingement

Steam impingement of the top row of tubes is a common problem. As Practical Example 4–3 illustrated, steam impingement can cause extensive degradation of tube material. This is a classic example of erosion-corrosion.

Tube Sheet Leaks

Leaks at the tube sheet, where tubes are rolled into the sheet, were rare during my years at the utility. Such leaks often appear as "weepers," where the leaks are small enough that an immediate shutdown is not required and an outage can be scheduled during a time of low electrical demand to repair the leak. Finding the leak can be the toughest problem. I have watched maintenance crews stretch plastic wrap across the inlet tube sheet in an attempt to detect small leaks. Theory says that the condenser vacuum will pull the wrap into the tube that is leaking, but as far as I could tell, more often than not this technique did not work. Dye-check procedures are often necessary to find very small leaks.

The inlet end of the condenser may be the location of another problem, that being erosion-corrosion of the tube inlets. The turbulence of the cooling water as it enters the tubes can strip metal ions from the surface, exposing fresh metal to the environment. Soft metals like Admiralty are particularly susceptible to this type of corrosion, and water velocities in Admiralty-tubed condensers should be limited to 7 linear feet-per-second (fps) or less. Other metals, while being more durable, may also suffer from this attack if inlet water velocities are too high or if the water contains excessive dissolved solids. The recommended water velocity for 90-10 copper-nickel is 10 fps for fresh water and 8 fps for seawater.

Failure of Tubes Due to Vibrational Effects

This attack is also known as fretting. Steam moving through the condenser can cause tubes to vibrate. The tubes then rub against the tube support plates, which generates erosion-corrosion of the tube walls. This attack is also known to occur in feedwater heaters.

Air In-Leakage

Excess air leakage into a condenser represents another potentially serious problem. Oxygen is quite reactive to copper and especially iron, so its ingress into condensate must be controlled to prevent corrosion and subsequent transport of corrosion products to the boiler.

Air enters a steam-generating system at points around the condenser due to the strong vacuum generated within. Prime spots for in-leakage include the expansion joint between the turbine and condenser, penetrations of heater drips lines into the condenser shell, turbine seals and explosion diaphragms, and condensate pump seals. Air in-leakage is virtually impossible to prevent, but under normal conditions the air-removal system can handle in-leakage. Oxygen not withdrawn by the air-removal system will dissolve in the condensate and travel downstream or will coat condenser tubes and reduce efficiency. When the air-removal system functions properly, condensate dissolved oxygen concentrations may be as low as 7 ppb. As several Practical Examples in chapter 6 illustrate, mechanical or structural failures can increase air in-leakage, sometimes very rapidly, and to a point where the air-removal system becomes overwhelmed. Condensate dissolved oxygen levels then rapidly increase.

Condensers are also a prime spot for air-inleakage to the feedwater system during times of shutdown. Also, if water is allowed to remain standing in the condenser, water vapor will creep into the turbine where it may combine with deposits on turbine blades and initiate corrosion. Moisture-induced sodium chloride corrosion of turbine blades is a well-recognized phenomenon.

Condensate Polisher

Condensate polishing is often not applied to low and medium-pressure boilers, as the benefits are not always cost effective. On the other hand, condensate polishing is an absolute requirement for once-through boilers and boilers on oxygenated treatment. The two common types of condensate polishers are deep-bed and powdered-resin. They may be designed for partial or full flow, but in either case have to handle a large water stream. Each type offers advantages and disadvantages. A powdered-resin polisher is comprised of filter elements to which a coat of finely-grained ion exchange resin is applied. The filter/resin combination performs the dual purpose of particulate removal and ion exchange. This type of resin is non-regenerable, so the resin must be replaced periodically, perhaps every month or two. One obvious advantage is that a powdered-resin unit does not require regeneration and so regenerant chemicals, feed systems, and chemical feed controls are not required. On the flip side is that the resin must be replaced frequently.

Deep-bed polishers are very similar to makeup demineralizers, in that the water passes through a bed of ion exchange resin, which removes ions while the bed slowly exhausts. The advantages and disadvantages are the reverse of those for powdered-resin units. Deep-bed polishers will also filter particulates, which eventually begins to restrict flow through the bed. In many if not most cases, a deep-bed polisher is not run to resin exhaustion but rather to a pressure-differential setpoint, at which time the bed is taken out of service for regeneration.

Polishers are designed to protect the boiler from contamination, but can introduce impurities if not properly operated. This is especially true with deep-bed polishers. In some designs, regeneration takes place directly within the exchange vessel. This offers the possibility that acid or caustic regenerant could be accidentally introduced to the condensate system.

Condensate/Feedwater System

The condensate/feedwater system conveys condensate from the hotwell to the boiler. A typical utility condensate/feedwater system includes low-pressure heaters, a deaerating heater, high-pressure heaters, and often an economizer. Low-pressure boilers, such as those at light or medium industrial plants, may only have a deaerator within this network.

The condensate portion of the system is defined as that piping and equipment from the condenser to the deaerator. The feedwater system is the portion from the DA to the boiler. Bleed steam from the turbine is the heating source for the feedwater heaters. The systematic and progressive heating of the condensate and feedwater increases boiler efficiency, but this has limits. Eight heaters including the deaerator are just about the maximum for any system.

Materials in the condensate/feedwater system include carbon-steel piping and stainless steel and/or copper-alloy feedwater heater tubes. Chemistry in these systems is designed around corrosion prevention. Oxygen is one of the chief corrodents. During normal operation, carbon-steel piping develops a thin layer of magnetite, Fe_3O_4, which protects the base metal.

$$3Fe + 4H_2O \rightarrow Fe_3O_4 + H_2 \qquad [Eq.\ 4\text{--}7]$$

Similarly, the surface of copper alloys becomes covered with a layer of protective cuprous oxide (Cu_2O). Excess oxygen converts the magnetite to rust (Fe_2O_3) and the cuprous oxide to cupric oxide (CuO). Neither of these compounds is protective. Rust is non-adherent and will detach from the pipe walls. The copper reacts with ammonia in the same manner as that shown in Equation 4–6 and produces the soluble complex $Cu(NH_3)_4^{+2}$. This soluble compound is washed away by the flowing feedwater stream.

One of the primary problems caused by feedwater system corrosion, besides degradation of materials, is generation of corrosion products that travel downstream to the boiler, where the high heat fluxes cause deposition on boiler tube walls. Iron oxides in general form the primary deposits on boiler tubes, and are most responsible for the need to chemically clean the boiler. Copper may also plate out on boiler tube walls, and in high-pressure units (2400 psig and above) will carry over with steam and precipitate on turbine blades. Just a few pounds of copper in a turbine can lead to significant degradation of performance. Feedwater corrosion products may also be directly introduced to the turbine if the feedwater serves as the attemperating medium to main and reheat steam.

Because oxygen control is so important, I wish to discuss an important piece of equipment within the condensate/feedwater network, that being the deaerator (DA). A DA consists of a steam scrubbing vessel and storage tank. Spray, tray or combination spray/tray deaerators are the norm. Condensate is introduced into the top of the scrubbing compartment and flows downward through the trays or is sprayed into steam injected into the compartment. The steam scrubs the condensate and raises its temperature to within a few degrees of saturation. The process liberates dissolved gases including oxygen, whose solubility decreases with increasing temperature (Table 4–2). The liberated gases are vented through the top of the deaerator. Proper deaerator performance is dependent upon several factors including correct alignment of the trays, evenly distributed condensate and steam flow, and sufficient venting. Any upset in these conditions could allow excess oxygen to pass on to the feedwater system.

Solubility of Oxygen vs. Temperature

Temperature (°F)	Oxygen Concentration (cc per liter)
30	10.1
50	7.8
70	6.2
90	5.1
110	4.4
130	3.8
150	3.1
170	2.4
190	1.5
210	0.1

Table 4–2 *Solubility vs. Temperature*

Oxygenated Treatment

With all the talk of oxygen control in steam-generating systems, it may come as a surprise to some readers that an increasingly-popular feedwater treatment program is based on the direct injection of oxygen into the condensate/feedwater system. This is known as oxygenated treatment, and we will look at it in the next chapter. We will also look at flow-assisted-corrosion (FAC). This phenomenon is causing great concern in the utility industry. FAC occurs when strongly-reducing waters strip the protective magnetite layer away from carbon-steel piping. Primary corrosion locations include elbows in feedwater lines and short-radius bends in economizers. A number of catastrophic failures have occurred in the last decade, resulting in several fatalities. FAC is also of concern in heat recovery steam generators, as we shall see.

The other major corrosion-influencing parameter in feedwater systems is pH. Both copper alloys and carbon steel exhibit minimal corrosion at an alkaline pH, although the optimum pH range for each is slightly different. We will look at pH control ranges in greater detail in chapter 5, when we examine chemical treatment programs. For the time being it is sufficient to know that most feedwater systems are treated with a chemical oxygen scavenger and a pH-conditioning agent to protect the piping and heat exchangers from corrosion. Hydrazine (N_2H_4) was once the scavenger of choice, although it has fallen into some disfavor due to environmental and safety concerns. Ammonia is a common pH conditioner, although amines molecules are also very popular.

While these chemicals do combat corrosion, they may also cause problems if not well controlled. Excess hydrazine will break down at temperatures above 400°F to form ammonia:

$$3N_2H_4 \rightarrow N_2\uparrow + 4NH_3 \qquad [Eq.\ 4\text{--}8]$$

This reaction can increase ammonia concentrations beyond those produced by ammonia or amines fed for pH control, and can increase corrosion of copper alloys. Organic amines may break down in the boiler to form organic acids and carbon dioxide, which then carry over to the turbine and cause corrosion. There still remains a lot of debate on the pros and cons of amines versus ammonia for feedwater pH control.

Boilers

The boiler is of course the heart of the steam generating system. Conditions in a boiler are obviously very harsh, and here chemistry problems are most magnified. As a start to this section let's take a brief look at different boiler designs.

Drum-Type Boilers. The most popular boilers at industrial facilities and electric utilities are of the drum variety. They are so called because of the presence of a steam drum, which gives the boiler unique operating characteristics. Large industrial and utility drum boilers must be erected in the field, but smaller boilers are often shipped and installed as package units.

Package Drum Boilers. Many industrial plants generate steam only for process use or to drive small turbines. For these applications, pre-constructed, package boilers are often sufficient. Figures 4–3 through 4–5 illustrate the simplified circuitry of three of the most common package boiler designs, the "A", "D", and "O" types. Natural gas or oil is the principal fuel for these boilers. Steaming rates typically range from 40,000 to 200,000 pounds per hour, with 100,000 pounds per hour being very common. As the illustrations indicate, each of the boilers contains one or more mud drums. The mud drum(s) serves as a collection site for precipitates formed by chemical treatment programs.

Field-Erected Drum Units. For the higher steam quantities or pressures needed at electric utilities and industrial plants that produce significant amounts of power, field-erected units are the norm. Custom-built units utilize long, vertical waterwall tubes, and are thus much larger and taller than package units. Although gas and oil-fired boilers are not uncommon, the predominant fuel is coal. Several boiler designs have evolved over the years. During the heyday of large boiler construction, the most common types included Cyclone; wall-fired; and pulverized-coal (PC), tangentially-fired boilers. The steam drum is located near the top of the unit, and the boiler either contains a mud drum or lower headers for collection of precipitates. High temperature superheater and reheater tubes are hung vertically just before or in the gas passage from the boiler, although some horizontally-arranged superheater and reheater tubes may be located further downstream in the gas passage. Larger units are often equipped with an economizer that is placed in the gas passage near the low-temperature superheaters and reheaters.

CHAPTER 4: FUNDAMENTALS OF STEAM GENERATION
AND SOURCES OF CONTAMINATION

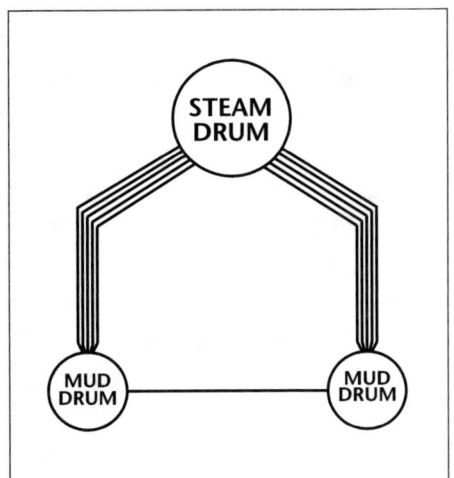

Fig. 4-3 *"A" Boiler Design*

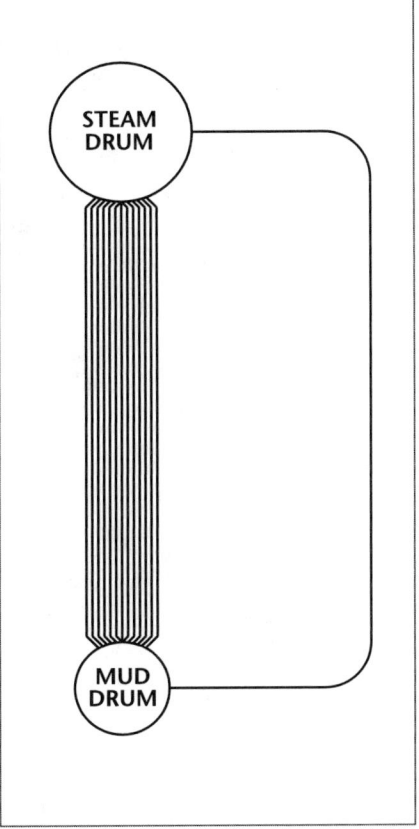

Fig. 4-4 *"D" Boiler Design*

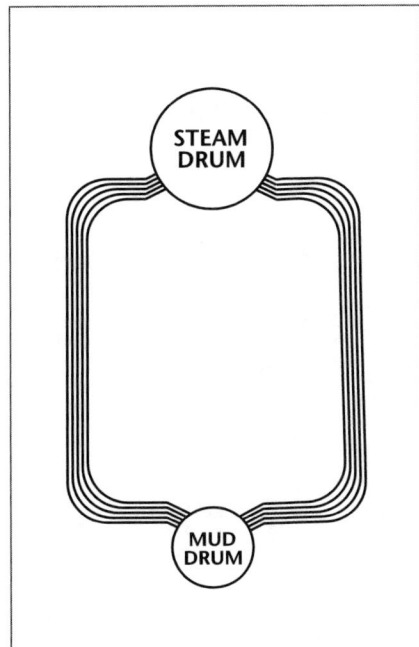

Fig. 4-5 *"O" Boiler Design*

Heat Recovery Steam Generators

Due to the changing nature of the electric utility industry, and more stringent environmental regulations, other power generating technologies are gaining acceptance and recognition. The most popular choice nowadays is combined-cycle generation, where a portion of the power is produced in a combustion turbine, and whose exhaust gases produce steam in an HRSG to drive a steam turbine or feed an industrial process (co-generation). The features of combined-cycle operation that have proven most appealing include unit efficiencies approaching 60 percent, cost, the ability to fire with natural gas or oil as a backup, and relative ease of installation. HRSGs are unique because the steam generator may contain two or three successively higher pressure circuits, each with its own drum. Water chemistry varies from one circuit to another and makes water treatment somewhat complex. Several HRSG designs are available including units with vertical waterwall tubes and horizontal waterwall tubes. The former, which more closely resemble conventional boilers are most popular.

The aspect of drum-type units that so greatly affects chemistry is the continual circulation of water from the steam drum through relatively cooler downcomers to the waterwall tubes and back (Figure 4–6). Steam

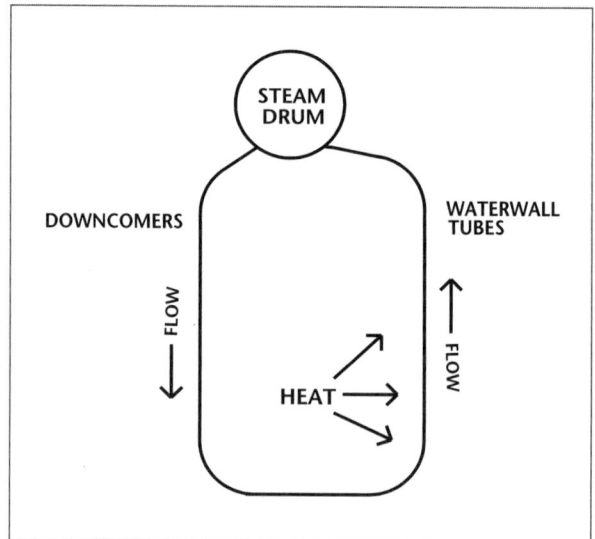

Fig. 4–6   *Water Circulation in a Drum Boiler*

CHAPTER 4: FUNDAMENTALS OF STEAM GENERATION
AND SOURCES OF CONTAMINATION

produced in the waterwalls is collected in the upper portion of the drum for distribution. The circulation/evaporation process concentrates solids in the boiler water. Drum units are susceptible to deposition and corrosion from contaminants introduced via condenser leaks, condensate return upsets, makeup system malfunctions, corrosion of downstream feedwater piping, and corrosion generated by shutdowns and startups. Let's take a look at some of these mechanisms. Shutdown and startup issues are so important, that I included a separate section for it at the end of this chapter.

Iron Oxide Deposition

As we learned in the discussion on feedwater chemistry, carbon steel develops a tightly bound layer of magnetite when the unit is placed in operation. The same process occurs with boiler waterwall tubes. Over the passage of time, however, the protective film becomes overlayed with more porous deposits of magnetite, which consist of iron-oxide corrosion products transported from the feedwater system. These porous deposits have much lower heat transfer coefficients than the tube metal, and can eventually reduce boiler efficiency and cause localized overheating of tubes. The iron oxide may become intermixed with other compounds including copper, hardness salts, and silica. Besides restricting heat transfer, deposits may also directly contribute to corrosion of the tubes.

Consider Figure 4–7, which shows a porous tube deposit. Water penetrates the deposit through various channels. As the water approaches the tube surface, temperatures increase. The water boils off, leaving other

Fig. 4–7   *Chimney Boiling in a Deposit*
Illustration by Alyssa Buecker

species behind. A common form of caustic attack occurs by this mechanism. Boiler water is normally maintained in an alkaline range (pH of 9 to 11 depending on boiler pressure and type) to minimize general corrosion. This pH range is moderately basic, and small amounts of caustic alkalinity may be present in the bulk boiler water. As water boils off in the deposit, sodium hydroxide is left behind. Concentrations may rise to levels many times that in the bulk boiler water. The concentrated NaOH attacks the boiler metal and protective magnetite film via the following reactions:

$$Fe + 2NaOH \rightarrow Na_2FeO_2 + H_2\uparrow \qquad [Eq.\ 4\text{--}9]$$

$$Fe_3O_4 + 4NaOH \rightarrow 2NaFeO_2 + Na_2FeO_2 + 2H_2O \qquad [Eq.\ 4\text{--}10]$$

The localized attack may cause tube damage within a relatively short period of time. One particular difficulty with iron oxide deposition is that the particles tend to precipitate on the hot side of the tubes. This is the worst location, as high temperatures increase the potential for corrosion. Under-deposit corrosion mechanisms will proceed more rapidly at higher temperatures.

Practical Example 4–2 outlined the rapidity in which a condenser leak can upset boiler water chemistry. The following equations illustrate several of the most common problems caused by cooling water ingress to the boiler.

$$Ca^{+2} + 2HCO_3^- \rightarrow CaCO_3\downarrow + CO_2\uparrow + H_2O \qquad [Eq.\ 4\text{--}11]$$

$$Ca^{+2}\ (or\ Mg^{+2}) + SiO_3^{-2} \rightarrow CaSiO_3\downarrow\ (or\ MgSiO_3\downarrow) \qquad [Eq.\ 4\text{--}12]$$

$$MgCl_2 + 2H_2O \rightarrow Mg(OH)_2\downarrow + 2HCl \qquad [Eq.\ 4\text{--}13]$$

Equations 4–11 and 4–12 are typical scale-forming reactions. Table 4–3 lists the most common deposits found in boilers. The effect of such deposition on heat transfer is outlined in Table 4–4, which shows thermal conductivities for various boiler metals and scales. Even a relatively thin deposit layer will significantly reduce heat transfer, and a boiler must be fired harder to achieve the same level of steam production. This in turn can lead to overheating of the boiler tubes, which will shorten tube life.

Common Boiler Deposits

Common Name	Chemical Formula
Acmite	$Na_2O \cdot Fe_2O_3 \cdot 4SiO_2$
Analcite	$Na_2O \cdot Al_2O_3 \cdot 4SiO_2 \cdot 2H_2O$
Anhydrite	$CaSO_4$
Aragonite	$CaCO_3$
Brucite	$Mg(OH)_2$
Calcite	$CaCO_3$
Cancrinite	$4Na_2O \cdot CaO \cdot 4Al_2O_3 \cdot 2CO_2 \cdot 9SiO_2 \cdot 3H_2O$
Hematite	Fe_2O_3
Hydroxyapatite	$Ca_{10}(OH)_2(PO_4)_6$
Magnetite	Fe_3O_4
Noselite	$4Na_2O \cdot 3Al_2O_3 \cdot 6SiO_2 \cdot SO_4$
Pectolite	$Na_2O \cdot 4CaO \cdot 6SiO_2 \cdot H_2O$
Quartz	SiO_2
Serpentine	$3MgO \cdot 2SiO_2 \cdot 2H_2O$
Thenardite	Na_2SO_4
Wallastonite	$CaSiO_3$
Xonotlite	$5CaO \cdot 5SiO_2 \cdot H_2O$

Table 4-3 *Most Common Boiler Deposits*
Source: Betz Handbook of Industrial Water Conditioning, Ninth Edition. BetzDearborn is a division of Hercules, Inc.

Effect of Boiler Deposits on Heat Transfer

Material	Thermal Conductivity (BTU/ft^2-hr-°F-in)
Carbon Steel	360
Magnetite	20
Calcium Carbonate	7
Porous Silica	0.6

Table 4-4 *Effect of Boiler Deposits on Heat Transfer*

Scaling of course can also lead to under-deposit corrosion, where tube failure may be much more rapid.

The reaction shown in Equation 4–13 is representative of that which caused the huge drop in boiler water pH outlined in Practical Example 4–3. Not only does acid cause general corrosion, but the reaction of acid with iron generates hydrogen, which can lead to hydrogen damage of the tubes. In this mechanism, hydrogen gas molecules, which are very small, penetrate into the metal wall where they then react with carbon atoms in the steel to generate methane (CH_4):

$$2H_2 + Fe_3C \rightarrow 3Fe + CH_4\uparrow \qquad [Eq.\ 4\text{–}14]$$

Formation of the very voluminous methane molecules causes cracking in the steel, greatly weakening its strength. Hydrogen damage is very troublesome because it cannot be easily detected. After hydrogen damage has occurred, the plant staff may replace tubes only to find that other tubes continue to rupture.

Chemistry upsets and corrosion have also been known to occur due to problems with chemical treatment programs. Phosphate hideout is one well-known phenomenon. I have covered it and other treatment-related problems in Chapter 5.

Chemistry in Heat Recovery Steam Generators

Combined-cycle power generation represents the future of electricity production. For the new generation of chemists and operators who will have to run these boilers, an understanding of HRSG chemistry is almost essential. Several features of HRSGs stand out over conventional units. Some of the most important are the following:

- HRSGs may have two or three steam-generating boilers, each at successively higher pressures.
- The low-pressure boiler may serve as the feedwater source for the higher-pressure circuit.
- Heat fluxes are usually quite a bit lower. Some combined-cycle units are equipped with supplemental burners, however, to augment heating of the HRSG.

Chapter 4: Fundamentals of Steam Generation and Sources of Contamination

- The deaerator may be integral to the low-pressure boiler.
- Water flow rates in HRSGs are generally higher.
- Combined-cycle units are typically started up and shut down many times during a year. Due to the ease of firing, the unit may be started up much more quickly than a coal-fired boiler.
- Makeup water requirements can be quite large for co-generation units, where part of the steam may be consumed on an industrial process.

The first thought one might have in reading this list is that lower heat fluxes mean less severe corrosion and scale formation than in a standard boiler. For a base-loaded HRSG unit this would generally be true. However, frequent startups and shutdowns drastically alter the situation. Startups and shutdowns cause problems in any boiler, but in HRSGs particularly they induce the following:

- Thermal stress and corrosion fatigue. This condition is exacerbated by quick starts.
- Increased deposit formation in the boiler.
- Increased carryover to the steam turbine.
- Increased possibility of flow-assisted-corrosion.
- Fluctuating drum levels at startup.
- Fluctuating feedwater and boiler water temperatures at startup.
- Increased dissolved oxygen levels and subsequent corrosion in deaerators, economizers, boilers, and superheaters.
- Carbonic acid corrosion during layups.

For these reasons, water chemistry is just as, if not more, important in HRSGs than in conventional boilers. Design aspects add an extra complexity to chemistry issues. Multiple boiler circuits may require separate chemical feed systems. One common HRSG design has the low-pressure boiler directly feeding the higher-pressure circuits. This requires very pure water in the low-pressure boiler with a zero dissolved solids treatment program. We will look at HRSG chemistry control methods in chapter 5.

Once-Through Steam Generation

Once-through steam generation is a fairly popular design for utility boilers, and is the only method to produce steam at or above supercritical pressure. All of the incoming feedwater is converted to steam in the waterwall tubes, which is then collected in steam headers for distribution. Because water/steam separation is not possible, the incoming feedwater has to be virtually free of dissolved solids. Boiler water chemistry is essentially one and the same with feedwater chemistry. An absolute requirement for once-through units is condensate polishing.

Drum Boiler Chemistry and Operation and the Effects on Steam Chemistry

Steam produced in the boiling process is never completely pure, and even at its best contains trace amounts of impurities. A primary mechanism by which solids are transferred to steam is known as carryover. Carryover is influenced by the following:

- Virtually all solids are at least slightly soluble in steam. Many solids become more soluble as boiler pressure increases and some, particularly silica, carry over extensively as a vapor.

- Even with the best steam separating devices, moisture droplets still enter the steam. This is known as mechanical carryover.

- Chemicals introduced for feedwater or boiler water treatment may increase carryover.

Let's look at some of these mechanisms in more detail.

Mechanical Carryover

During the steam generation process in drum boilers, water droplets become entrained in the steam. The droplets are of course boiler water.

CHAPTER 4: FUNDAMENTALS OF STEAM GENERATION
AND SOURCES OF CONTAMINATION

If the droplets are allowed to remain, they will introduce dissolved solids to the superheater, reheater, and turbine. Boiler drums, therefore, contain steam separating devices to remove the bulk of the entrained moisture. Figure 4–8 shows a simplified diagram of a common steam separating scheme. The separation equipment relies upon the difference in density between steam and water. In the design shown, steam first passes through cyclonic separators, which throw water droplets to the walls where they drain back into the drum. The steam then passes through chevron-like scrubbers that remove additional water.

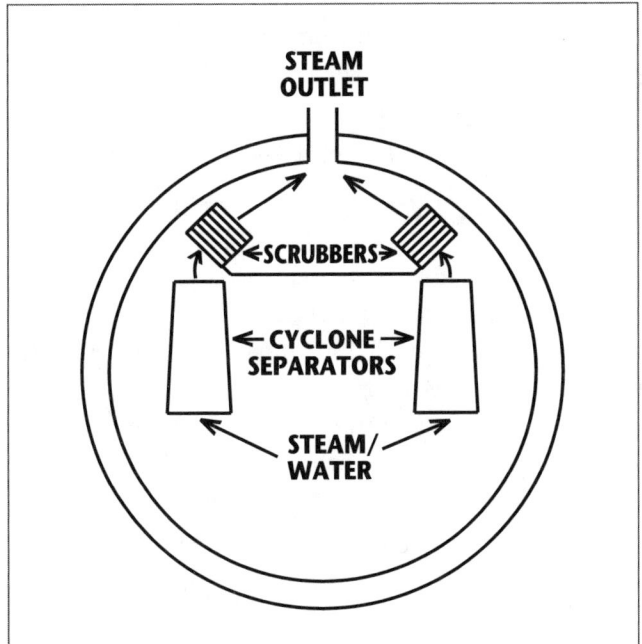

Fig. 4–8 *Common Steam Separating Scheme*

Even these devices will not prevent steam contamination if boiler water chemistry or operating conditions are not properly controlled. The most common factors that influence mechanical carryover include foaming, priming, improper drum water level control, poor drum design or operation outside of boiler design limits.

Priming is the introduction of slugs of water to the steam. This frequently occurs during large and rapid load swings, when the demand for steam changes quickly. Priming can also be caused by poor drum level control, wherein high drum levels overload the steam separating equipment. Foaming is caused by excessive concentrations of solids in the boiler water. As the foam bubbles burst, they introduce excessive quantities of water to the steam separating devices, which may not be able to totally remove the added moisture. As we saw in Practical Example 4–1, foaming may be more prevalent at industrial facilities, where makeup treatment systems are less sophisticated and condensate polishing may be minimal or non-existent.

Inadequate drum design will influence all of the problems mentioned above. One of the most important factors in drum design is that sufficient space be provided between the operating water level and the steam separation devices. If this space is too small, slugs of water or excess water vapor can enter the steam separating devices. Operation of the boiler at higher than rated capacity can also overload the steam separating internals.

Vaporous Carryover

Silica is the most well known vaporous carryover product. It is also one of the most troublesome for several reasons. First, it can form tenacious deposits on turbine blades, which can often only be totally removed by sandblasting or chemical cleaning. Second, carryover potential greatly increases with increasing boiler pressure. Third, silica can easily enter a boiler system. It is the most weakly held ion on a makeup system anion exchanger, and will break through first. Another not-uncommon source is leakage of collodial silica through ion exchange systems, which we learned about in chapter 3. I have also heard of silica being introduced when contractors grit-blasted turbine blades using a silica-based material. Residuals of the material that entered the condenser lay in the hotwell and gradually leaked silica into the system. Yet another example was given to me by a good friend at a midwestern utility. After a scheduled unit outage, silica levels remained high in the boiler for weeks afterwards. He checked for condenser tube leaks, collodial silica intrusion, and every other mechanism he could think of. Finally, he discovered that during the outage maintenance personnel used a different sealant when closing the door on the deaerator storage tank manhole. Guess what? The sealant contained silica. Once they removed the material, the problem disappeared.

Solids Introduction by Contaminated Attemperator Water

Steam temperature control is important for efficient operation of a boiler. Although steam temperature is a function of boiler design; changes in operation, load, and firing patterns require some form of external steam temperature control, especially for high-pressure units. This is often accomplished by direct injection of feedwater into the superheater and reheater sections of the unit. Poor feedwater chemistry leads to direct contamination of the superheater, reheater, and turbine.

Superheater Exfoliation

The stresses experienced by steam-generating equipment during shutdown, start-up, and cycling duty increase the mechanical degradation of the equipment. This occurs in the superheater and reheater, where the stress and harsh environment cause exfoliation of the magnetite layer. These particles may then pass to the turbine where they can cause erosion of the turbine blades. This effect, known as solid particle erosion, is most commonly found in the high-pressure end of the turbine where linear velocities are highest.

Sometimes, corrosion of afterboiler components may occur due to excess oxygen in the steam. A prime location for this corrosion is the crossover line between the high-pressure and low-pressure ends of the turbine, where any slight amount of moisture helps generate the corrosion reaction.

Steam Chemistry

The turbine is designed to extract every available bit of usable heat from the steam. Ideally, this means taking the steam from its maximum superheat at the turbine inlet to saturated conditions at the outlet. The condenser then removes the latent heat and converts the steam to condensate. Practically, however, a small portion of the steam begins to condense in the low-pressure end of the turbine. This is known as the saturation line, and its onset is influenced by boiler load and other factors. The section of the turbine at or near the saturation line is a critical location for corrosion and fouling, as we shall see.

The most common turbine deposits or corrodents include silica, magnetite, copper, sodium chloride, hydroxide, and phosphate, organic acids, and ammonium chloride and sulfate. Figure 4–9 shows the relative solubility of these and other compounds in steam. As the diagram indicates, the solubility of all these compounds decreases with decreasing steam pressure. So, as steam passes through the turbine and pressure falls off, the compounds collect on the turbine blades. When excessive contamination

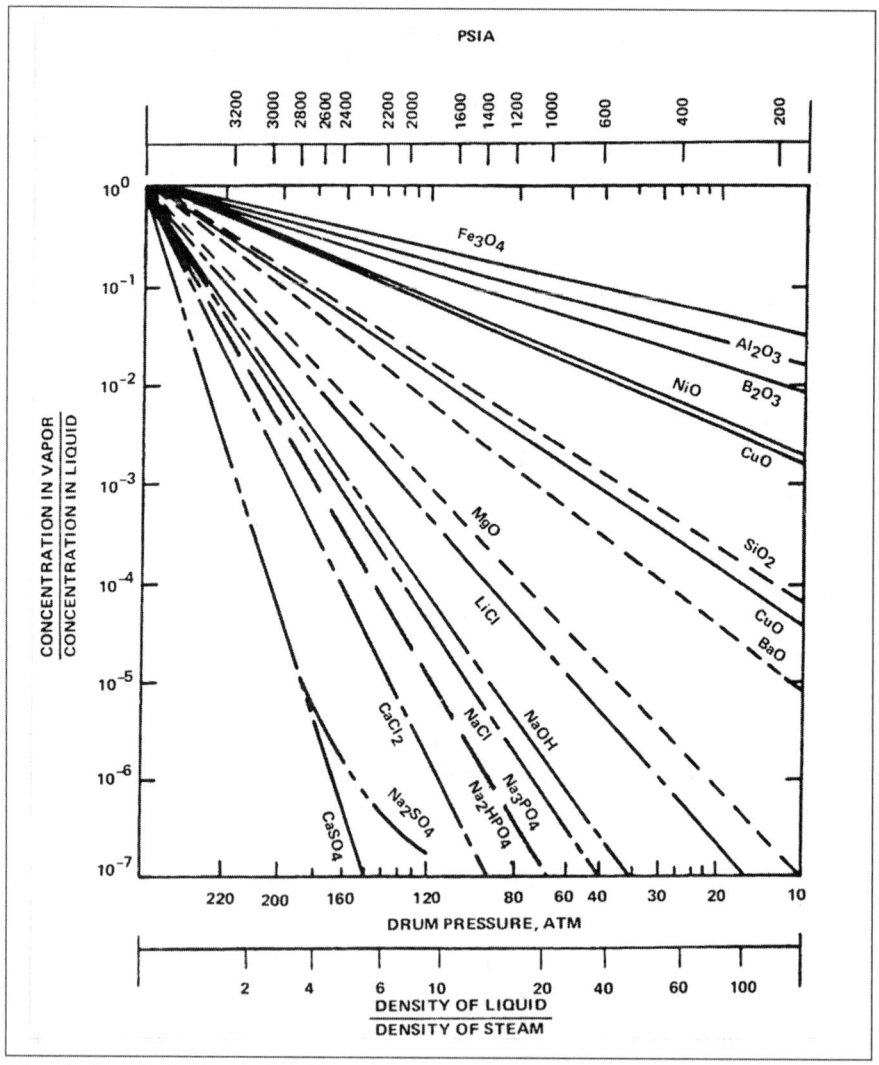

Fig. 4–9 *Relative Solubility of Compounds in Steam*

of boiler water or feedwater occurs, contaminants may precipitate throughout the turbine. (See Practical Example 4–7). But under normal circumstances, the solubility of each compound dictates the general location where the contaminant drops out of solution. Many compounds concentrate in the low-pressure end of the turbine.

Figure 4–10 illustrates the characteristic deposition pattern in the low-pressure stage. Several interesting patterns are evident. First, sodium compounds and salts tend to precipitate just ahead of the saturation line. These include the chloride and sulfate salts. The compounds can initiate stress corrosion cracking, pitting, and corrosion fatigue. Second, on either side of the saturation line are zones labeled "Drying on Hot Surfaces." Moisture formation may allow contaminants to deposit in these zones, but then a subsequent fluctuation in temperature dries out the deposits and allows them to remain. The area between the saturation line and the 4% moisture line (the Wilson line) is another critical location. Here solubility of impurities is lowest, but not enough moisture is available to wash deposits from the blades. Finally, 6% moisture is the point where formation of carbonic and organic acids due to the carryover of CO_2 or

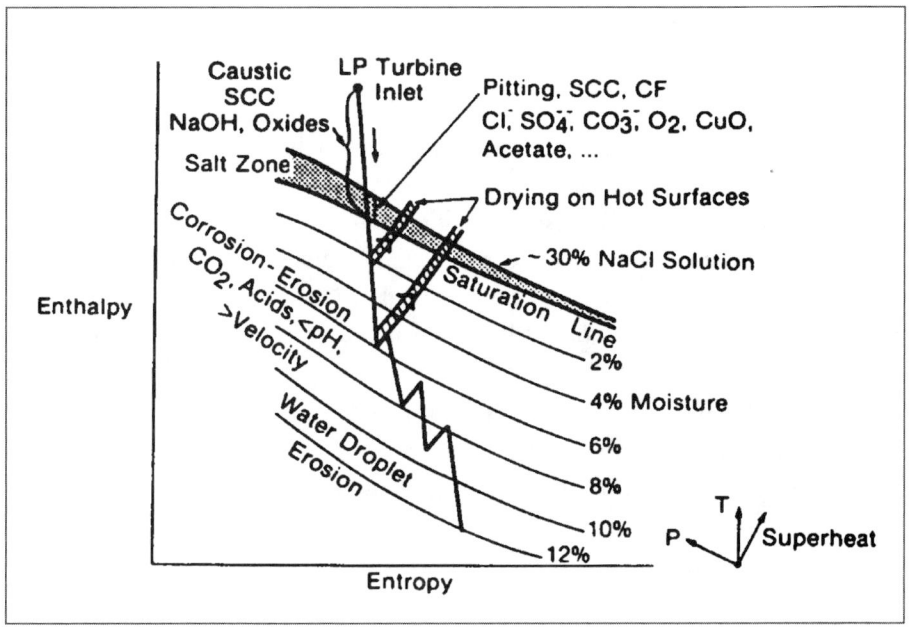

Fig. 4–10 *Deposition Pattern in Low Pressure End of a Turbine*
Source: EPRI

organics to the turbine. Deposition and corrosion are not entirely limited to the low-pressure turbine. The following sections outline details of specific contaminants.

Contaminants

Copper. Copper deposition in turbines generally affects units that operate at 2400 psig and above. Copper tends to precipitate in the high-pressure stages of the turbine, where it forms copper oxide or metallic copper deposits. Over time, the deposits will degrade turbine performance, and many cases of unit derating due to copper deposition have been recorded. Just a few pounds of copper may cause a double-digit loss of megawatt capacity. This phenomenon has been under close scrutiny in recent years, as chemists strive to understand the mechanisms and magnitude of the problem.

Sodium Hydroxide. Sodium hydroxide is soluble in water and steam, and exists as a dissolved compound in both phases. Sodium hydroxide is a primary culprit in stress corrosion cracking (SCC). Turbine blades are naturally under stress because of operating conditions, and thus are susceptible to SCC.

Chloride and Sulfate. Several compounds, including chloride and sulfate salts, concentrate in a narrow region defined as the salt zone. Chlorides induce pitting in stainless steels, which are the material of construction for turbine blades and rotors. Chlorides and sulfates also cause corrosion fatigue and SCC of turbine blades. These two corrosion mechanisms alone cost U.S. utilities hundreds of millions of dollars per year in replacement costs and lost power.

Chloride and sulfate carryover into steam is influenced by the ammonia concentration in the boiler water, which allows the compounds to carry over as ammonium salts. Higher ammonia levels increase chloride and sulfate carryover.

Although chlorides and sulfates are themselves harmful to turbine blades, the salts can become acidic and cause additional corrosion. In fact, acid corrosion of turbine blades has been observed to be much higher in

AVT units than phosphate-treated units. Evidence indicates that sodium phosphate carryover helps to neutralize acidic compounds that may form on the turbine components.

Iron Oxides. Iron oxides transported by high velocity steam cause solid particle erosion (SPE) of the turbine blades. Turbine screens remove some of these particles, but small particles may pass through the screens and erode the edges of the high-pressure turbine blades. Exfoliation of superheater piping generates many of these iron oxide particles.

Silica. Silica will coat turbine blades and gradually cause a decrease in turbine performance. Although it is sometimes possible to water wash silica deposits from the blades, silica frequently forms a hard scale that is difficult to remove. Chemical cleaning may be required. Silica is typically the limiting factor as load is increased after a unit startup. Table 4–5 shows the most common silicate deposits in turbines. Amorphous silica is generally the primary material.

Sodium Phosphates. Sodium phosphates will carry over and deposit on intermediate- and low-pressure turbine blades and in superheaters and reheaters. If the deposits are not too heavy this can actually present an advantage, as the phosphates help neutralize acids formed by other carryover products. However, excessive phosphate carryover is detrimental. More than one case has been reported of reheater tube failures due to sodium phosphate accumulation in reheater U-bends. During one reheater tubing at my former utility, my supervisor and I inspected a number of reheater U-bends. Each contained a mixture of exfoliated magnetite mixed with tri-sodium phosphate. The failures occurred at these U-bends because the deposits had so restricted heat transfer that the tubes failed due to long-term overheating.

Organics. Breakdown of organics in the boiler or steam lines introduces organic acids and carbon dioxide to the turbine. These acids can cause corrosion in the low-pressure blades. Debate still rages as to how severely organic amines and oxygen scavenengers influence turbine corrosion. Research groups like EPRI recommend hydrazine and ammonia as feed-

Common Silicate Deposits

Chemical Formula	Name
SiO_2	Silica
Na_2SiO_3	Sodium Metasilicate
$Na_2SiO_3 \cdot 5H_2O_3$	Sodium Metasilicate Pentahydrate
$Na_2SiO_3 \cdot 9H_2O_3$	Sodium Metasilicate Nonahydrate
$NaAlSiO_4$	Sodium Aluminum Silicate
$NaAlSi_3O_{12}(OH)$	Sodium Aluminum Silicate Hydroxide
$Na_4Al_6SO_4(SiO_4)_8$	Sodium Aluminum Sulfate Silicate
$NaFeSi_2O_6$	Sodium Iron Silicate
$Na_3[Cl_6(AlSiO_4)_6]$	Sodium Chlorohexaaluminum Silicate
$KAlSi_3O_8$	Potassium Aluminum Silicate
$KNa_3(AlSiO_4)_6$	Potassium Trisodium Aluminum Silicate
$Mg_6[(OH)_8Si_4O_{10}]$	Magnesium Octahydride Silicate
$Mg_3Si_4O_{10}(OH)_2$	Magnesium Silicate Hydrate
$Ca_2Si_2O_4$	Calcium Silicate
$Ca_2Al_2Si_3O_{10}(OH)$	Calcium Aluminum Silicate Hydroxide
$3Al_2O_3 \cdot 4Na_2O \cdot 6SiO_2 \cdot SO_4$	Noselite
$(Fe,Mg)_7Si_3O_{22}(OH)_2$	Iron Magnesium Hydroxide Silicate
$Na_8Al_3Si_6O_{24}MoO_4$	Sodium Aluminum Molybdenum Oxide Silicate

Table 4–5 *Common Silica Deposits in Turbines*
Source: Betz Handbook of Industrial Water Conditioning, Ninth Edition. BetzDearborn is a division of Hercules, Inc.

water treatment chemicals whenever possible. The major water treatment chemical manufacturers disagree, and suggest that amines and organic oxygen scavengers provide a better and safer alternative.

Practical Example 4-7

This problem occurred in a 300 MW, 2600 psig, coal-fired unit. The plant has an open recirculating cooling system. Makeup to the tower comes from a saline river supply, and at four cycles of concentration, cooling water TDS concentrations average 20,000 ppm. The unit had been originally equipped with an on-line monitoring system, however, most of the instruments were no longer reliable. The lab relied primarily on grab sampling. On the day of the upset, a condenser tube failed around 5:00 p.m., two hours after the regular set of 3:00 p.m. grab sample readings. A chemist who arrived for routine shift duties at 11:00 p.m., found that the boiler water pH contained extremely high concentrations of impurities and that the boiler water pH was near 2! The operators, having received the 3:00 p.m. report and not being familiar with water chemistry, assumed that all was well. They did notice that the turbine was losing power, but did not attribute this to a chemistry problem. Since a regular unit outage was scheduled a day later, the unit remained on-line for one more day. The unit came off for the outage just as turbine performance began a very accelerated decline. Upon removal of the turbine casing, plant personnel discovered that the turbine blades from front to back were covered with salt deposits. Cleanup of the boiler and turbine added an extra 30 days to the outage. Total costs for turbine cleaning and overhaul, boiler cleaning, and purchased power came to around five million dollars. The utility immediately went to work planning and designing a new on-line water chemistry monitoring system.

Startups and Shutdowns

Startups and shutdowns of units are tougher on system components than almost any other mechanism. Heating and cooling subject the unit to thermal and mechanical stresses. We have already seen that this can lead to exfoliation corrosion in superheaters and reheaters. Frequent cycling may also initiate corrosion fatigue in some areas. Drum levels and water temperatures may fluctuate during startups, and this can lead to carryover of contaminants to steam and thermal stress. Finally, startups and shutdowns or even load swings can play havoc with boiler chemistry. It is easy to understand why boiler experts always recommended base-load operation when possible. Unfortunately from a chemistry standpoint, the current trend is just the opposite. The growth of combustion turbine power technology is the big factor in this trend. Simple cycle turbines make it easy for a utility to follow load swings. Because combustion turbines have short startup times, the utility can put the turbine on as load rises during the day and take it off again at night. This philosophy is carrying over to combined-cycle operation.

When a boiler cools during shutdown, everything shrinks including water level. The boiler develops a vacuum that will draw air into the unit at many spots. Locations include the superheater, boiler drum, deaerator, and feedwater heaters. Oxygen combined with water is extremely corrosive. Rust will form in superheaters and drums, while feedwater heater tubes and internals will suffer exfoliation and dealloying due to oxygen contamination. Air intrusion into the steam generating system also brings in carbon dioxide, which reacts with water to form carbonic acid and lowers pH.

The stresses caused by shutdowns and startups also release corrosion products into the system. Dissolved and suspended iron concentrations in boiler water have been known to reach almost 100,000 ppb immediately after a unit startup. This can introduce more iron oxide particulates to the hot boiler tubes than many hours of regular operation. In fact, one of the criteria for determining the need to chemically clean a boiler is the number of shutdowns and startups. In the first few hours after a startup, wet steam in the turbine will wash off silica deposits, which then show up in the feedwater and boiler only to be reintroduced to the turbine.

Cycling also impacts boiler water treatment chemistry. For many years, one of the most common treatment schemes in boilers has been coordinated or congruent phosphate. During normal operation, a portion of the phosphate compounds may precipitate (hide out) on tube walls. If boiler load is steady, the plant chemists can often make adjustments for hideout and maintain chemistry within standard guidelines. Load swings, and especially shutdowns and startups, influence the deposition mechanism, which can lead to wild changes in boiler water chemistry. It is easy to see that these aspects make water treatment of HRSGs a particularly complex issue.

Conclusion

This chapter was designed to inform the reader about the general chemistry in steam generating systems and the problems that can occur. The next two chapters discuss treatment programs, control guidelines, and monitoring techniques.

Learning Aide 4-1

COMPOSITION OF MATERIALS COMMONLY USED IN STEAM GENERATION SYSTEMS

A variety of materials are used in power plant construction. Heat exchanger tubes may be carbon steel, stainless steel, copper-alloy, or titanium. Condensate/feedwater piping and boiler waterwall tubes are usually made of carbon steel. Tubes in higher temperature regions of the boiler such as the superheater are fabricated of higher grades of carbon steel and sometimes even stainless steel. Turbine blades are stainless steel.

This learning aide outlines the composition of several of the most commonly-used materials in steam-generating systems. Subsequent discussions of corrosion are more understandable if one has an idea of the metal under attack.

For carbon and low-alloy steels, the American Iron and Steel Institute (AISI) and Society for Automotive Engineers (SAE) developed the following guidelines, where XX refers to the percent carbon.

- 10XX — Plain carbon steels.
- 11XX — Carbon steels with higher than normal sulfur for easier machining.
- 13XX — Contain 1.75% manganese (Mn).

- 23XX — Contain 3% nickel (Ni).
- 25XX — Contain 5% Ni.
- 31XX — Contain 1% Ni and some chromium (Cr).
- 33XX — Contain 3% Ni and some chromium (Cr).
- 40XX — Contain 0.25% molybdenum (Mo).
- 41XX — Contain 0.25% Mo and 1% Cr.
- 43XX — Contain 0.25% Mo, 0.75% Cr, and 1.75% Ni.
- 50XX — Contain 0.3% Cr.
- 51XX — Contain 1.0% Cr.
- 52XX — Contain 1.5% Cr.
- 61XX — Contain 1% Cr and 0.15% vanadium (V).
- 86XX — Contain 0.5% Cr, 0.5% Ni, and 0.2% Mo.
- 92XX — Contains 2% silicon (Si).

The other minor elements in these steels include manganese (0.7%), sulfur (0.040% max.), phosphorous (0.040% max.), and silicon (0.3%).

Maximum Operating Temperatures of Common Boiler Materials

Material Composition	Maximum Operating Temperature (°F)
Carbon Steel	850
1.25% Cr – 0.5% Mo	1100
2.25% Cr – 1% Mo	1200
3% Cr – 1% Mo	1200
9% Cr – 1% Mo	1200
Stainless Steel (18% Cr – 8% Ni)	1500

Table 4–6 *Heat Tolerance of Common Boiler Materials*

CHAPTER 4: FUNDAMENTALS OF STEAM GENERATION
AND SOURCES OF CONTAMINATION

Plain carbon steel is the common material for boiler waterwall tubes. The steel contains no chromium or molybdenum. This material is not suitable at temperatures much above 850°F, so other alloys are required in higher temperature areas of the boiler. The chrome-moly or even stainless steels serve as replacements. Table 4–7 shows the compositions of some of the other materials used in power plant applications.

Composition of Iron- and Nickel-based Alloys Commonly Used in the Steam Generation Industry

Material	Fe	C	Si	Mn	Cr	Ni	Mo	Other	Use
Austenitic Stainless Steel									
304L	Bal.	0.03	0.6	1.0	18.0	8.0	—		HX tubes
316	Bal.	0.05	0.6	1.0	18.0	13.0	2.1		HX tubes
316L	Bal.	0.03	0.6	1.0	17.5	13.0	2.1		HX tubes
Martensitic-Ferritic Steel									
Type 410	Bal.	—	—	0.15	12.5	0.75	—		Turbines
Type 430	Bal.	—	—	0.12	17.5	0.75	—		Turbines
6% Moly–Stainless Steel									
254 SMO	Bal.	0.02	0.5	0.4	20.0	18.0	2.1	0.2N	HX tubes
Nickel-based Alloys									
C276		0.01	0.08	0.6	15.5	Bal.	16.0	3.5W	FGDS
C22		0.01	0.04	0.4	22.0	Bal.	13.0	2.5W	FGDS

Composition of Common Copper-based Alloys

Admiralty	90-10 Cu-Ni	70-30 Cu-Ni	Muntz Metal
Cu – 71%	Cu – Bal.	Cu – Bal.	Cu – 61%
Zn – Bal.	Ni – 10%	Ni – 30%	Zn – Bal.
Sn – 1%	Fe – 1.25%	Fe – 0.55%	
As, Sb, or P – 0.06%	Mn – 0.3%	Mn – 0.5%	

Table 4–7 *Composition of Iron- and Nickel-based Alloys Used in Power Plant Applications*

It is important to note the functions of the major alloying elements, especially in the stainless steels.

Chromium. Chromium is what makes stainless steel stainless. At 12% concentrations or greater, chromium causes the steel to form an oxide layer that resists corrosion. Many of the commercial stainless steels contain 18% chromium. Chromium adds other properties to steel, but the stainless aspect is by far the most important.

Carbon. Carbon acts as a hardening element in steel. Concentrations must be carefully controlled, as excess carbon will impart brittleness to steel. Cast iron is a good example.

Nickel. Nickel in 8% concentrations or higher causes steel to remain in the austenitic (face-centered-cubic) crystalline structure at ambient temperature. This structure imparts better corrosion resistance in many applications. Nickel increases toughness and ductility of steel, and improves corrosion resistance.

Molybdenum. Molybdenum adds strength and also increases the passive film strength. Stainless steels containing molybdenum are more resistant to pitting and chloride attack. For example, the chloride limit for 304 SS is around 500 ppm, but for 316L SS it rises to about 3,000 ppm.

Manganese and Silicon. Both improve oxidation resistance. Manganese is also a desulfurizer.

Learning Aide 4-2

EFFECT OF THE CONDENSATION PROCESS ON STEAM TURBINE EFFICIENCY

For anyone who might have a slight interest in thermodynamics, the following illustration indicates why steam condensation is so important to plant efficiency.

When water converts to steam in a boiler, several phenomena occur. First, the water heats to the boiling point. Then, a phase transition takes place as the water changes to steam. This phase change requires a fairly large amount of energy known as the latent heat of vaporization. If water is only given enough heat to make the phase change, then the steam exists in a saturated state. It might be suitable for providing heat to a process but not for driving a turbine. Most boilers are equipped with a superheater to add heat to saturated steam. Superheating by itself increases because the steam can perform more work in the turbine. However, a great deal of extra efficiency is gained by condensing the steam after its passage through the turbine.

Consider what happens in a condenser. Ideally, when steam leaves the turbine it has used all of its available heat for work and is at a saturated condition. The condenser of course converts the steam to condensate, which produces a very strong vacuum due to the enormous decrease in volume. This is very important thermodynamically.

Let's look at the following hypothetical situation.

- Steam, at a pressure of 1,000 psig and temperature of 950°F enters a turbine.
- The turbine is adiabatic and reversible, meaning no energy is lost to the surroundings and entropy does not change.
- The turbine has no bleed lines for feedwater heating.

The steam tables (by the way the steam tables are being updated right now) show that the enthalpy of the turbine inlet steam is 1477 BTU/lbm and the entropy is 1.6325 BTU/lbm °R. Because the turbine is adiabatic and reversible, the entropy of the exhaust steam is the same as the entropy of the throttle steam since no heat is transferred during the reversible expansion.

Accordingly, we can calculate the enthalpy of the exhaust steam for various conditions. In the hypothetical turbine, if the steam were taken from the turbine exhaust at atmospheric pressure (14.7 psia), its enthalpy is about 1081 BTU/lbm. Thus, 396 BTU/lbm of heat is available for work. The enthalpy of the condensate at these conditions is 180 BTU/lbm. Turbine efficiency may be calculated by the following equation:

$$\%EF = \frac{h_i - h_f}{h_i - h_c} \times 100 \qquad [Eq.\ 4\text{--}15]$$

Where: h_i = enthalpy of turbine inlet steam
h_f = enthalpy of turbine outlet steam
h_c = enthalpy of condensate

Thus the efficiency of our hypothetical turbine discharging to atmospheric pressure becomes

$$\frac{1477 - 1081}{1477 - 180} \times 100 = 33\% \qquad [Eq.\ 4\text{--}16]$$

If, however, all incoming steam exhausts into a vacuum, efficiency improves noticeably. With relatively cool circulating water, absolute condenser pressures can reach as low as 0.5" Hg (0.3 psia). The enthalpy of steam exhausting into these conditions is near 843 BTU/lbm, and the condensate enthalpy is around 27 BTU/lbm. The turbine efficiency becomes

$$\frac{1477 - 843}{1477 - 27} \times 100 = 44\% \qquad [Eq.\ 4\text{--}17]$$

It is clearly evident that the condensation process greatly improves efficiency. This example is exaggerated, as no turbine is reversible. Also, some steam is usually extracted from the turbine for feedwater heating. But, the example still shows that condenser performance can dramatically affect unit efficiency. In winter weather it is possible to cool the condensate too much, in which a small amount of efficiency is lost due to the extra heat required to raise the condensate temperature in the feedwater heaters. This is known as condensate subcooling.

CHEMICAL TREATMENT PROGRAMS FOR STEAM GENERATING SYSTEMS | 5

Now that we have examined the basics of steam generation and the sources and effects of water chemistry upsets on boiler operation, we will turn to chemical treatment programs that protect steam generating systems. This chapter outlines the most popular treatment methods and provides chemistry control guidelines. A good treatment program can be worth its weight in gold, or perhaps more accurately iron, copper, and the other metals that make up boiler tubes, turbine blades, and other steam generating materials. Also of great importance are layup treatment methods. Chemical treatment starts in the condensate/feedwater system.

Condensate/Feedwater Treatment

The primary purpose for treating the boiler feedwater system is to minimize corrosion of piping, feedwater heater tubes, and the economizer. Figure 5–1 illustrates one of the most common corrosion mechanisms. The most critical factors in preventing feedwater system corrosion are control of dissolved oxygen and pH. We will look at both issues.

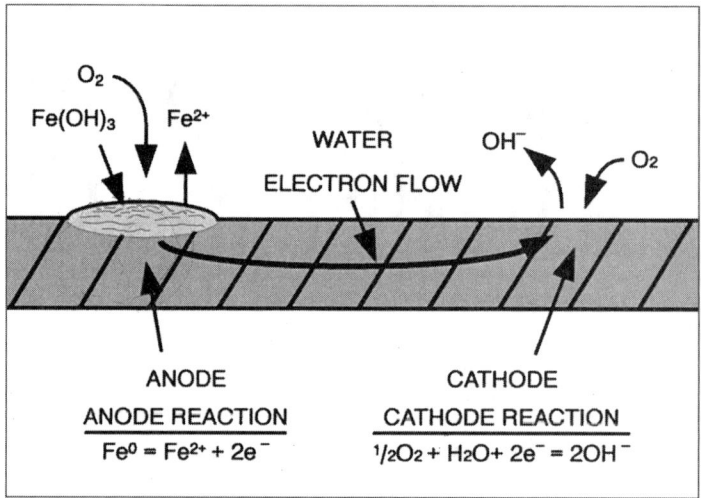

Fig. 5–1   *A Common Iron Corrosion Mechanism*
Source: Betz Handbook of Industrial Water Conditioning, Ninth Edition. BetzDearborn is a division of Hercules, Inc.

Control of Dissolved Oxygen

In chapter 4 we learned that air may enter a steam generating system at many locations around the steam surface condenser or in aerated makeup. Ingress of air, and the principal corrodent oxygen, is controlled by both mechanical and chemical methods, including these:

- Mechanical removal of gases in the condenser
- Mechanical removal of gases by a deaerator located in the feedwater system
- Injection of an oxygen scavenger into the feedwater

The details of the air-removal system were outlined in Chapter 4. The Heat Exchange Institute (HEI) has established a design guideline of 7 ppb dissolved oxygen in condensate from a properly operating condenser. Other guidelines suggest that 20 ppb is a more practical figure.

Chapter 5: Chemical Treatment Programs
for Steam Generating Systems

Downstream from the condenser, the deaerator (DA) provides additional mechanical removal of dissolved gases. As condensate is sprayed or flows through trays downward into the deaerator scrubbing vessel, the rising steam heats the condensate and liberates gases. Table 4-2 already illustrated the effect of temperature on dissolved oxygen concentrations, and a properly operating deaerator should reduce dissolved oxygen concentrations to 7 ppb.

Chemical Control of Oxygen

Mechanical methods are not the sole means employed to control dissolved oxygen; and in many steam generating systems, chemicals act as a supplement to mechanical air-removal devices. One of the first practical chemicals to be employed was sodium sulfite (Na_2SO_3). Sodium sulfite reacts with oxygen to produce sodium sulfate:

$$2Na_2SO_3 + O_2 \rightarrow 2Na_2SO_4 \qquad [Eq.\ 5\text{--}1]$$

Sodium sulfite has a molecular weight almost four times greater than that of molecular oxygen, and reacts in a 2 to 1 molar ratio, so theoretically 8 ppm of Na_2SO_3 are needed to remove one ppm of oxygen. Sulfite residuals are often maintained at 30 ppm or higher to provide adequate protection.

The primary advantages of sodium sulfite are that it is a common and inexpensive chemical, is non-toxic, and can be used to treat boiler water in which the steam is extracted for food processing or other FDA-regulated applications. Sodium sulfite is primarily used in low-pressure industrial boilers (<600 psig), because it adds too many dissolved solids to high-pressure boiler water. Also, in boilers that operate above 900 psig, sodium sulfite will thermally decompose to produce hydrogen sulfide (H_2S) and sulfur dioxide (SO_2), both of which are quite corrosive.

$$Na_2SO_3 + H_2O \rightarrow 2NaOH + SO_2\uparrow \qquad [Eq.\ 5\text{--}2]$$

$$4Na_2SO_3 + 2H_2O \rightarrow 3Na_2SO_4 + 3NaOH + H_2S\uparrow \qquad [Eq.\ 5\text{--}3]$$

For utility and industrial boilers that operate at pressures above 900 psig, alternative chemicals are more suitable for oxygen scavenging. The workhorse for many years has been hydrazine (N_2H_4), which reacts with oxygen as follows:

$$N_2H_4 + O_2 \rightarrow 2H_2O + N_2\uparrow \qquad [Eq.\ 5\text{-}4]$$

Hydrazine proved advantageous because it does not add any dissolved solids to the feedwater, it reacts with oxygen in a one-to-one weight ratio, and it is supplied in liquid form at 35% concentration.

A primary benefit of hydrazine is that it will passivate oxidized areas of piping and tube materials as follows:

$$N_2H_4 + 6Fe_2O_3 \rightarrow 4Fe_3O_4 + N_2\uparrow + H_2O \qquad [Eq.\ 5\text{-}5]$$

$$N_2H_4 + 4CuO \rightarrow 2Cu_2O + N_2\uparrow + 2H_2O \qquad [Eq.\ 5\text{-}6]$$

One part of N_2H_4 will theoretically passivate 30 parts of Fe_2O_3 and 10 parts of CuO. Hydrazine residuals are maintained at much lower levels than sodium sulfite, typically in a range from 20 to 100 part-per-billion (ppb). (This is provided that the hydrazine does not remove all traces of dissolved oxygen. A slight oxygen residual is needed to prevent flow-accelerated-corrosion, which is detailed later in this chapter.)

Hydrazine performance is greatly enhanced by increased temperature and pH, but precautions must be taken to prevent overfeed of the chemical, as excess hydrazine will begin to decompose at temperatures above 400°F to form ammonia:

$$3N_2H_4 \rightarrow N_2\uparrow + 4NH_3 \qquad [Eq.\ 5\text{-}7]$$

This can interfere with the supplemental ammonia or amine compounds that are used for pH control. Breakdown of excess hydrazine can raise ammonia concentrations to levels that significantly increase copper-alloy corrosion.

Hydrazine does not react rapidly at the temperatures found on the condensate side of a condensate/feedwater system. In fact, tests by EPRI showed that reaction rates are several times greater at a temperature of

200°C than 100°C. So, the recommended injection point for hydrazine once was the deaerator storage tank or boiler feed pump suction. The addition of catalysts to hydrazine greatly improved its performance at lower temperatures. Cobalt is one such catalyst, and hydroquinone ($C_6H_6O_2$, Figure 5–2), which is itself an oxygen scavenger, is another. The improved performance of catalyzed hydrazine convinced some utility personnel to switch the injection point from the deaerator storage tank/boiler feed pump suction to the condensate pump discharge. This provides protection to the condensate piping and low-pressure feedwater heaters. The greatest drawback to this method is that some of the hydrazine is removed in the deaerator.

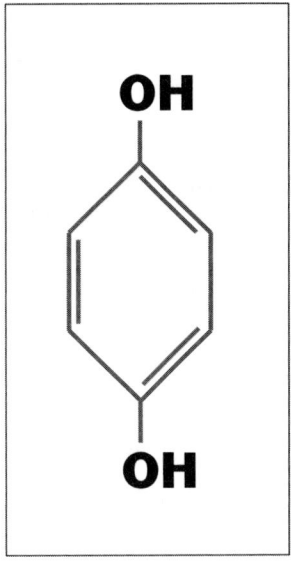

Fig. 5–2 *Hydroquinone*

Given the simplified chemistry of hydra-zine, it would appear to be the ideal oxygen scavenger. Unfortunately, hydrazine is considered to be a potential carcinogen and is now registered as a hazardous compound. Handling procedures have become much more stringent. This difficulty has in part led to the development of other treatment chemicals. Many of the major water treatment companies provide alternative oxygen scavengers, the principal ones being hydroquinone, carbohydrazide (N_4H_6CO, Figure 5–3), and methyl ethyl ketoxime (C_4H_9NOH, Figure 5–4). All of these products also passivate metals.

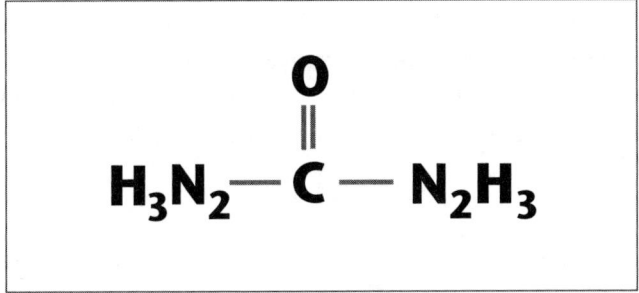

Fig. 5–3 *Carbohrdrazide*

Carbohydrazide is a derivative of hydrazine and actually breaks down to hydrazine and carbon dioxide as it is heated in the feedwater system.

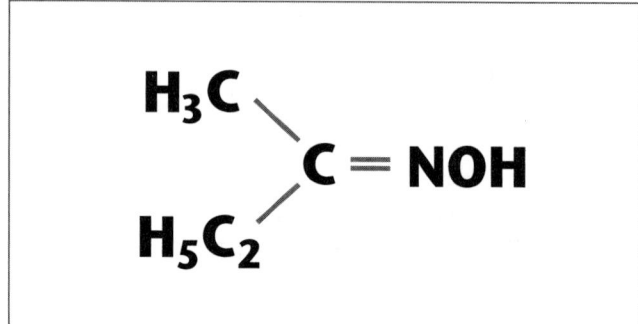

Fig. 5–4   *Methyl Ethyl Ketoxime*

The chief advantage of carbohydrazide is that it is not considered to be hazardous like hydrazine, and can be handled more easily. A drawback is that carbon dioxide evolves when carbohydrazide reacts with oxygen or passivates metals:

$$N_4H_6CO + 2O_2 \rightarrow N_2\uparrow + 2H_2O + 3CO_2 \qquad [Eq.\ 5\text{–}8]$$

$$N_4H_6CO + 12Fe_2O_3 \rightarrow 8Fe_3O_4 + 2N_2\uparrow + 3H_2O + CO_2 \qquad [Eq.\ 5\text{–}9]$$

$$N_4H_6CO + 8CuO \rightarrow 4Cu_2O + 2N_2\uparrow + 3H_2O + CO_2 \qquad [Eq.\ 5\text{–}10]$$

Carbohydrazide has performed well in some units, and is reported to be a better metal passivator than uncatalyzed hydrazine at low temperatures. Some facility operators use carbohydrazide for wet layups of boilers due to its effectiveness at ambient temperatures.

Hydroquinone is a benzene-derivative compound that has been used as a stand-alone oxygen scavenger and as a catalyst for hydrazine. Hydroquinone reacts with oxygen to produce benzoquinone.

$$C_6H_6O_2 + \tfrac{1}{2}O_2 \rightarrow C_6H_4O_2 + H_2O \qquad [Eq.\ 5\text{–}9]$$

Due to its organic structure, hydroquinone and its oxidized products can break down to a variety of organic acids and carbon dioxide in a boiler system.

Methyl ethyl ketoxime is an organic scavenger that reacts with oxygen to produce methyl ethyl ketone (MEK).

$$C_4H_9NOH + {}^1\!/_2 O_2 \rightarrow C_4H_9O + N_2O\uparrow + H_2O \qquad [Eq.\ 5\text{--}12]$$

Reported advantages of methyl ethyl ketoxime include a higher thermal stability than other scavengers, and a much higher distribution ratio in steam. The latter allows the compound to cycle through the system. However, methyl ethyl ketoxime and MEK can break down to smaller organics, carbon dioxide, and nitrogen compounds.

All of these products (hydrazine, carbohydrazide, hydroquinone, and methyl ethyl ketoxime) have performed well in some applications, and worse in others. A clear-cut choice is not always possible. For instance, one full-scale test of all four compounds showed that the relative scavenging or passivating properties of each chemical varied depending on location in the condensate/feedwater system. One single product did not outperform another in all locations or for all functions.

Some experts in the electric utility industry oppose the use of organic oxygen scavengers, as they generate organic acid and carbon dioxide decomposition products. This is a reasonable view, since organic acids and CO_2 are known to cause corrosion of turbine blades. But, the concentration at which the attack becomes serious is still being researched. Scientists are continuing their efforts to determine the species, quantities, and effects of organic acids and CO_2 in the turbine. The introduction of organics by oxygen scavengers may be slight in comparison to those introduced from pH-conditioning amines, which are typically maintained at higher concentrations in the feedwater than the oxygen scavenger.

Safe Hydrazine Systems

It is possible to set up a hydrazine feed system that does not expose workers to the compound. In one such system at a western utility, the hydrazine solution is supplied in portable, re-usable containers. Personnel connect each new container to a permanent metering pump and distribution line, which transport precise dosages of the chemical in neat form to the feedwater system. A variation of this concept is a system in which the primary feed tank supplies a closed day tank, which is vented to the outside atmosphere. The operator introduces a measured volume of hydrazine to the day tank, followed by dilution water. A metering pump feeds the solution into the system.

The portable feed concept is gaining great acceptance. The water treatment firm is responsible for delivering full tanks to the plant site and hauling away the empty vessels. Plant personnel do not have to deal with drums and related hazardous waste disposal requirements. The capacity of portable containers is typically within a range of 200 to 400 gallons. A method to minimize handling is to stack one portable feed container on top of a primary feed container. The top container is plumbed such that it drains into the bottom vessel. When the top container empties, it is replaced with a full vessel. This arrangement provides two distinct advantages. First, the bottom container can be permanently piped to the metering pump or connected to a day tank. Secondly, the top container drains completely without compromising the performance of the system. Very little residual chemical remains in the vessel.

Industrial Oxygen Scavenging

In the food, beverage, and pharmaceutical industries, process steam comes under FDA regulations. Besides sodium sulfite, an approved oxygen scavenger for these applications is erythorbic acid ($C_6H_8O_6$). Erythorbic acid is an isomer of Vitamin C, and is reported to react very speedily with oxygen. Erythorbic acid has been used in electric utility boilers, but like the other organic oxygen scavengers, it will break down at high temperatures to form organic acids and CO_2. Still, it is a good alternative in higher-pressure systems where sodium sulfite would cause too many problems.

Flow Assisted Corrosion

In chapter 4 we examined flow-assisted-corrosion, and the fact that conventional oxygen scavenging treatments greatly contribute to this phenomenon. FAC is also of concern in heat recovery steam generators (HRSGs), where flow rates and geometry influence FAC.

Several methods have been proposed to combat FAC in conventional steam-generating systems. One method is to back off on oxygen-scavenger feed, such that a 1 to 2 ppb oxygen residual remains in the feedwater. This helps preserve the magnetite layer without subjecting the system to corrosive levels of oxygen. To inhibit FAC of high-pressure heater drains, EPRI recommends that feedwater heater vents be closed. This, however, has caused problems at some utilities and must be evaluated on a case-by-case basis. And of course, oxygenated-treatment eliminates FAC entirely. For operators of systems with mixed metallurgy, finding a middle ground between optimum oxygen scavenging concentrations and prevention of FAC can be difficult. Copper-alloy corrosion is greatly minimized in a reducing environment generated by oxygen scavengers, but these are the conditions that cause FAC. The development of on-line oxidation-reduction-potential (ORP) techniques to monitor feedwater may offer a control method to optimize chemistry. More discussion about this possibility appears in chapter 6.

Condensate/Feedwater pH Control

Control of pH within the feedwater system is quite important. As we learned in Chapter 4, excursions of pH outside of a relatively narrow range induce corrosion, most notably in iron-based materials. Feedwater piping and heat exchanger tubes exhibit minimal corrosion at a mildly alkaline pH. For a feedwater system containing all steel metallurgy the optimum pH range is 9.0 to 9.6. Corrosion control in mixed-metallurgy systems is more complicated. Admiralty brass exhibits minimum corrosion within a pH range of 8.5 to 9. Copper-nickel alloys, particularly the 90-10 alloy, are most stable around a pH of 9.3. So, the question becomes "What is the best pH for a system containing carbon-steel piping and copper-alloy heat exchanger tubes?" A commonly-recommended pH range for these systems is 8.8 to 9.1.

Ammonia or organic amines are the pH-conditioning chemicals of choice. Table 5–1 lists the most popular compounds. The best choice depends upon several factors, including the pressure of the steam generating system, the complexity of the system being protected, and the characteristics of the additive. Of the compounds listed in the table, cyclohexylamine is most basic, while morpholine is much less basic. The basicity of diethylaminoethanol and ammonia lies in between. Thus, the chemist has a number of choices for finding a product that maintains condensate/feedwater pH within recommended guidelines.

Relative Basicity of Popular pH Conditioning Chemicals ($K_b \times 10^6$)

Compound	72°F	298°F	338°F
Ammonia	20.6	6.9	4.6
Cyclohexalamine	489	61	32
Diethylaminoethanol	68	11.3	9.2
Morpholine	3.4	4.9	3.8

Table 5–1 *Most Popular pH Conditioning Compounds*
Source: *Betz Handbook of Industrial Water Conditioning, Ninth Edition. BetzDearborn is a division of Hercules, Inc.*

Amines will decompose to produce ammonia in feedwater. Whether the ammonia comes from direct ammonia feed or amine decomposition, a general rule-of-thumb recommends that ammonia concentrations be limited to 0.5 ppm in systems containing copper-alloy materials Even this level may be too high where frequent in-leakage of air is a problem.

In high-pressure utility boilers, where the steam is quite pure, decomposition of amines can potentially introduce unwanted organic acids and CO_2 to the turbine. For this reason, some experts recommend ammonia as the best pH-conditioning chemical. This is not universally accepted, however. The situation may be vastly different in an industrial boiler where bicarbonate decomposition causes heavy carryover of CO_2 to condensate lines. Amines can be a good product for neutralization of the carbon dioxide. A particular aspect of importance is the amine distribution ratio. This is the percentage of amine that carries over with the steam versus that which remains in the boiler water. The distribution ratio varies with the product and with the pressure of the boiler (Figure 5–5). Selection of the

CHAPTER 5: CHEMICAL TREATMENT PROGRAMS FOR STEAM GENERATING SYSTEMS

neutralizing chemical based on distribution ratio can be very important. If protection of afterboiler condensate lines is required, an amine with a high steam-to-liquid distribution ratio is best. Where corrosion prevention in the boiler is more critical, an appropriate neutralizing amine is one whose distribution ratio allows most of it to remain in the boiler water. Often, a blend of two amines will provide more comprehensive protection.

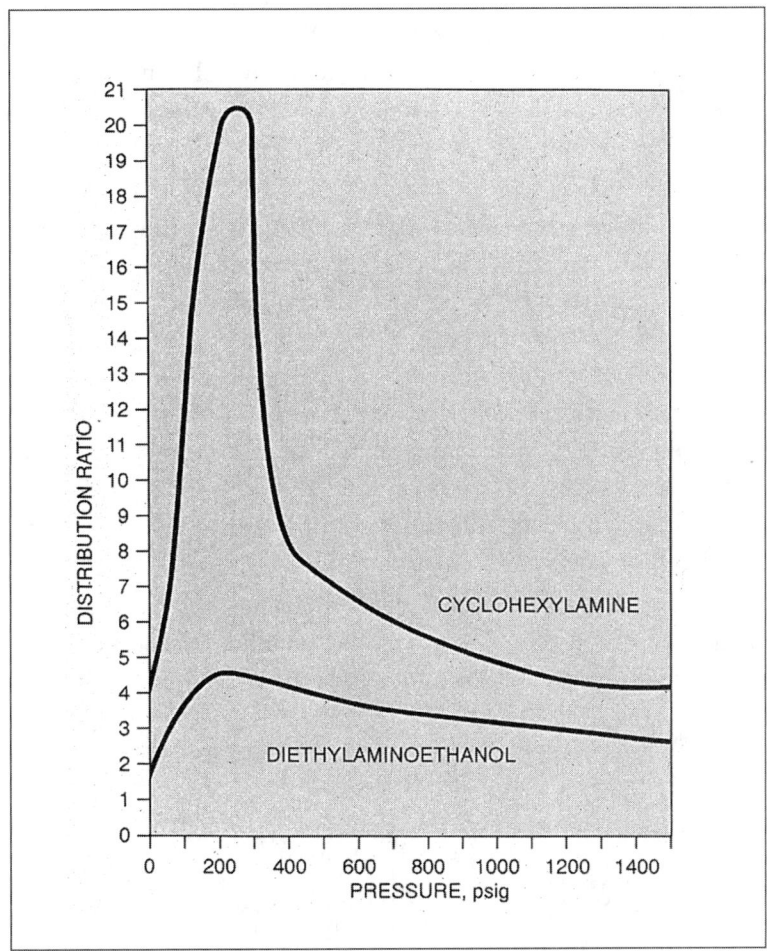

Fig. 5–5 *Amine Distribution Ratios vs. Temperature*
Source: Betz Handbook of Industrial Water Conditioning, Ninth Edition. BetzDearborn is a division of Hercules, Inc.

Oxygenated Treatment

In chapter 4 we looked at the concept of oxygenated treatment (OT). OT is a feedwater treatment that also serves to protect the boiler. In an OT program, oxygen is deliberately introduced to the condensate and feedwater system. Two variations of oxygenated treatment are most popular. In the first, oxygen is injected alone without any pH-conditioning chemicals. This program is known as neutral water treatment (NWT). More often, ammonia is injected for pH control. This is known as combined water treatment (CWT).

OT was developed in Germany about 30 years ago for replacement of all-volatile-treatment (AVT) in once-through steam generating units. The program was adopted by other European utilities, and now is gaining great acceptance at once-through utilities in the United States. The treatment requires the controlled injection of oxygen or hydrogen peroxide into the condensate/feedwater system. Figure 5–6 illustrates the normal injection points. In CWT programs, which are most common in the United States, oxygen is dosed to maintain a 50 to 150 ppb residual. Ammonia is added to raise the pH within a range of 8.0 to 8.5. Typically, 20 to 70 ppb of ammonia will produce this pH.

The chemistry of oxygenated treatment is very interesting and explains why the program has become popular. In conventional AVT units, even with very good chemistry programs, some iron oxides are still

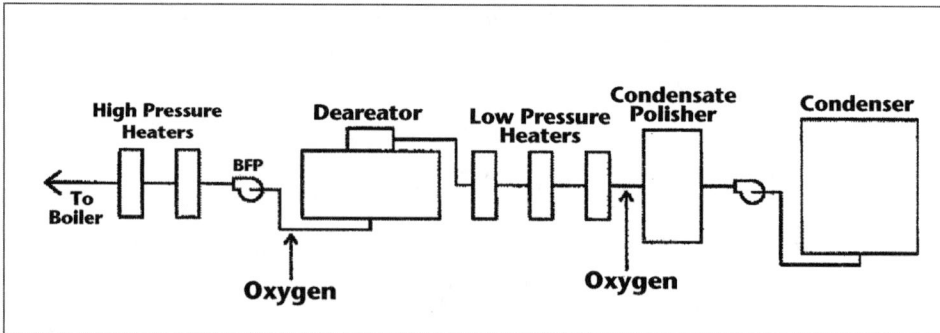

Fig. 5–6 *Normal Injection Points of H_2O_2 or into OT Systems*

generated in the condensate or feedwater system. It may be tough to keep iron concentrations below 10 ppb, which is just about the absolute maximum for high-pressure boilers. This is graphically illustrated in Figure 5–7, which shows the solubility of magnetite in ammonia-laden waters. With controlled oxygen injection however, the base layer of magnetite becomes covered and interspersed with an even tighter film of ferric oxide hydrate (FeOOH). (See Figure 5–8). This compact layer is more stable than magnetite and releases very little dissolved iron or suspended iron-oxide particles to the fluid. Some utilities that switched from AVT to OT have reported that dissolved feedwater iron concentrations, which were often 10 ppb or higher on AVT, dropped to as low as 1 to 2 ppb once the OT program was fully established.

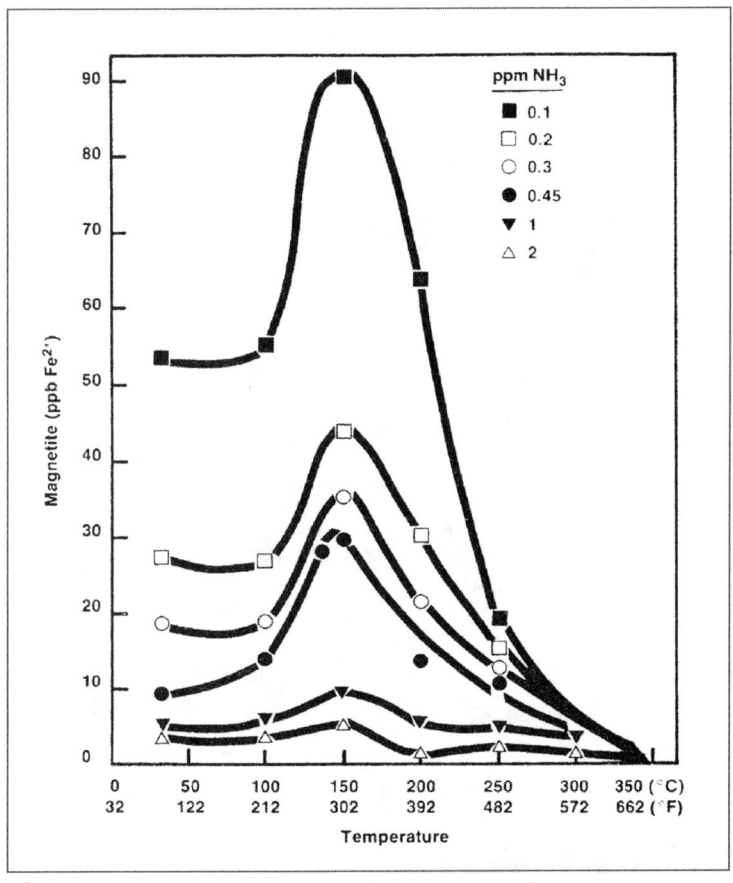

Fig. 5–7 *Solubility of Magnetite in Ammonia*

The keys to an OT program are controlled oxygen feed and high-purity makeup water. Makeup and feedwater must be pure in once-through units anyway, because contaminants would carry over directly with the steam. Deposits in OT units would be doubly detrimental due to the formation of differential oxygen cells, which would in turn cause under-deposit corrosion and pitting. Very pure feedwater in once-through units is usually a given, as these steam generating systems are always

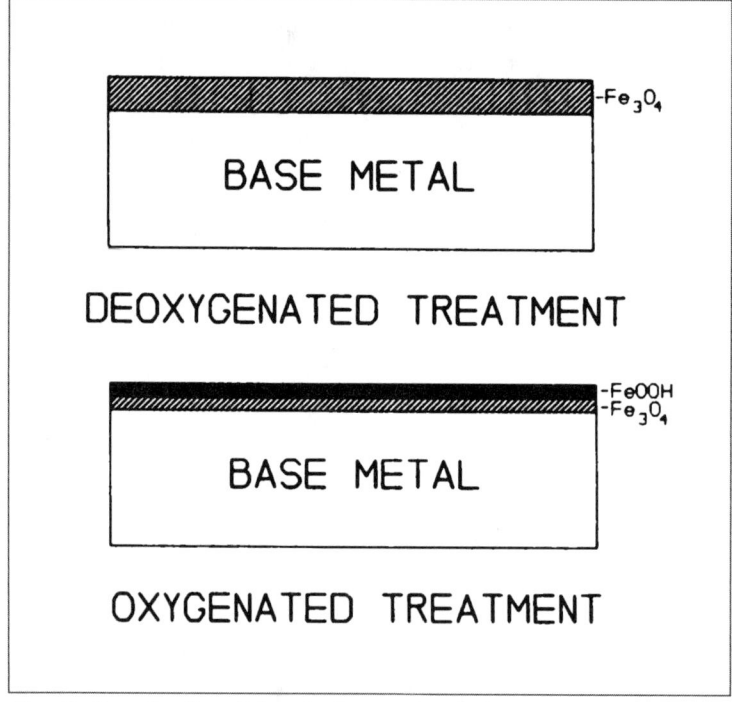

Fig. 5–8 *Surface Films in Deoxygenated vs. Oxygenated Treatments*

equipped with condensate polishers. OT cannot be used in systems that contain copper-alloy materials, as copper corrosion would be too great.

One of the major benefits of oxygenated treatment is that the programs eliminate flow-assisted-corrosion. The treatment constantly maintains an oxide layer on the feedwater pipe surfaces. Another positive about oxygenated treatment is that the feedwater pH is a unit or unit and a half lower

CHAPTER 5: CHEMICAL TREATMENT PROGRAMS
FOR STEAM GENERATING SYSTEMS

than that in AVT programs. This is due to the stability of the FeOOH film. The lower pH can greatly increase condensate polisher run lengths.

A couple of utilities are experimenting with oxygenated treatment in drum units. The results could be very interesting. The key point to remember is that OT will minimize transport of the chief boiler tube foulant, iron oxides. If other impurities are prevented from entering the system, tube fouling can be greatly reduced.

Chemistry Guidelines

We have examined feedwater chemistry and treatment methods, and are now prepared to look at boiler water programs. This is a good time to introduce some chemistry guidelines. These provide actual targets for feedwater and boiler water chemistry. Tables 5–3 through 5–5 are water chemistry guidelines from the American Society of Mechanical Engineers for drum boilers at various pressures. Table 5–2 is a summary of feedwater guidelines that have appeared in numerous papers at water treatment conferences. Other guidelines will appear throughout the remainder of this chapter. One important point to note about the guidelines is that feedwater and boiler water requirements become more stringent as pressures increase.

Feedwater Guidelines

Parameter	Volatile Treatment (Hydrazine/Ammonia) Mixed Metallurgy	All Ferrous Systems	Oxygenated Treatment
pH	8.8 – 9.1	9.2 – 9.6	8.0 – 8.5
Cation Conductivity (μs/cm)	<0.2	<0.2	<0.15
Iron (ppb)	<10	<5	<5
Copper (ppb)	<2	—	—
Dissolved Oxygen (ppb)	<5	1–10	30–150

Table 5–2 *Feedwater Guidelines*
Source: EPRI

SUGGESTED WATER CHEMISTRY LIMITS
INDUSTRIAL WATERTUBE, HIGH DUTY, PRIMARY FUEL FIRED, DRUM TYPE

Makeup water percentage: Up to 100% of feedwater
Conditions: Includes superheater, turbine drives, or process restriction on steam pubty
Saturated steam purity target: See tabulated values below.

Drum Operating Pressure [1][11]	psig 0–300 (MPa) (0–2.07)	301–450 (2.08–3.10)	451–600 (3.11–4.14)	601–750 (4.15–5.17)	751–900 (5.18–6.21)	901–1000 (6.22–6.89)	1001–1500 (6.90–10.34)	1501–2000 (10.35–13.79)
Feedwater [7]								
Dissolved oxygen ppm (mg/l) O_2 measured before chemical oxygen scavenger addition [8]	<0.007	<0.007	<0.007	<0.007	<0.007	<0.007	<0.007	<0.007
Total iron ppm (mg/l) Fe	≤0.1	≤0.05	≤0.03	≤0.025	≤0.02	≤0.02	≤0.01	≤0.01
Total copper ppm (mg/l) Cu	≤0.05	≤0.025	≤0.02	≤0.02	≤0.015	≤0.01	≤0.01	≤0.01
Total hardness ppm (mg/l) *	≤0.3	≤0.3	≤0.2	≤0.2	≤0.1	≤0.05	ND	ND
pH @ 25°C	8.3–10.0	8.3–10.0	8.3–10.0	8.3–10.0	8.3–10.0	8.8–9.6	8.8–9.6	8.8–9.6
Chemicals for preboiler system protection	NS	NS	NS	NS	NS	VAM	VAM	VAM
Nonvolatile TOC ppm (mg/l) C [6]	<1	<1	<0.5	<0.5	<0.5	<0.2	<0.2	<0.2
Oily matter ppm (mg/l)	<1	<1	<0.5	<0.5	<0.5	<0.2	<0.2	<0.2
Boiler Water								
Silica ppm (mg/l) SiO_2	≤150	≤90	≤40	≤30	≤20	≤8	≤2	≤1
Total alkalinity ppm (mg/l) *	<350 [3]	<300 [3]	<250 [3]	<200 [3]	<150 [3]	<100 [3]	NS [4]	NS [4]
Free OH alkalinity ppm (mg/l) * [2]	NS	NS	NS	NS	NS	NS	ND [4]	ND [4]
Specific conductance [12] μmhos/cm (μS/cm) 25°C without neutralization	5400–1100 [5]	4600–900 [5]	3800–800 [5]	1500–300 [5]	1200–200 [5]	1000-200 [5]	≤150	≤80
Total Dissolved Solids in Steam [9]								
TDS (maximum) ppm (mg/l)	1.0–0.2	1.0–0.2	1.0–0.2	0.5–0.1	0.5–0.1	0.5–0.1	0.1	0.1

* as $CaCO_3$ NS = not specified   ND = not detectable   VAM = use only volatile alkaline materials upstream of attemperation water source. [10]

Table 5–3 *ASME Water Chemistry Guidelines — 1*   Reproduced from Consensus on Operating Practices for the Control of Feedwater and Boiler Water Chemistry in Modern Industrial Boilers with Permission from the American Society of Mechanical Engineers.

CHAPTER 5: CHEMICAL TREATMENT PROGRAMS
FOR STEAM GENERATING SYSTEMS

Notes to Table 5-3

(1) With local heat fluxes $>1.5 \times 10^5$ Btu/hr/ft^2 (>473.2 kW/m^2), use values for at least the next higher pressure range.

(2) Minimum hydroxide alkalinity concentrations in boilers below 900 psig (6.21 MPa) must be individually specified by a qualified water treatment consultant with regard to silica solubility and other components of internal treatment.

(3) Maximum total alkalinity consistent with acceptable steam purity. If necessary, should override conductance as blow-down control parameter. If makeup is demineralized quality water and boiler operates at less than 1000 psig (6.89 MPa) drum pressure, the boiler water conductance should be that in table for 1001–1500 psig (6.9–10.34 MPa) range. In this case, the necessary continuous blowdown will usually keep these parameters below the tabulated maximum values. Alkalinity values in excess of 10% of specific conductance values may cause foaming.

(4) Not detectable in these cases refers to free sodium or potassium hydroxide alkalinity. Some small variable amount of total alkalinity will be present and measurable with the assumed congruent or coordinated phosphate-pH control or volatile treatment employed at these high pressure ranges.

(5) Maximum values are often not achievable without exceeding maximum total alkalinity values, especially in boilers below 900 psig (6.21 MPa) with >20% makeup of water whose total alkalinity is >20% of TDS naturally or after pretreatment by lime-soda, or sodium cycle ion exchange softening. Actual permissible conductance values to achieve any desired steam purity must be established for each case by careful steam purity measurements. Relationship between conductance and steam purity is affected by too many variables to allow its reduction to a simple list of tabulated values.

(6) Nonvolatile TOC is that organic carbon not intentionally added as part of the water treatment regime.

(7) Boilers below 900 psig (6.21 MPa) with large furnaces, large steam release space, and internal chelant, polymer, and/or antifoam treatment can sometimes tolerate higher levels of feedwater impurities than those in the table and still achieve adequate deposition control and steam purity. Removal of these impurities by external pretreatment is always a more positive solution. Alternatives must be evaluated as to practicality and economics in each individual case.

(8) Values in the table assume existence of a deaerator.

(9) Achievable steam purity depends on many variables, including boiler water total alkalinity and specific conductance as well as design of boiler steam drum internals and operating conditions (Note 5). Since boilers in this category require a relatively high degree of steam purity for protection of the superheaters and turbines, more stringent steam purity requirements such as process steam restrictions on individual chemical species or restrictions more stringent than 0.1 ppm (mg/l) TDS turbine steam purity must be addressed specifically.

(10) As a general rule, the requirements for attemperation spray water quality are the same as those for steam purity. In some cases boiler feedwater is suitable; however, frequently additional purification is required. In all cases the spray water should be obtained from a source that is free of deposit forming and corrosive chemicals such as sodium hydroxide, sodium sulfite, sodium phosphate, iron, and copper. The suggested limits for spray water quality are <30 ppb (μg/l) TDS maximum, <10 ppb (μg/l) Na maximum, <20 ppb (μg/l) SiO$_2$ maximum, and it should be essentially oxygen free.

(11) Low pressure boilers frequently use feedwater that is suitable for use in higher pressure boilers. In these cases the boiler water chemistry limits should be based on the pressure range that is most consistent with the feedwater quality.

(12) Conversion from ppm (mg/l) TDS values in the ABMA standards [12] used a factor of 0.65.

Table 5-3 ASME Water Chemistry Guidelines — 1 *Reproduced from Consensus on Operating Practices for the Control of Feedwater and Boiler Water Chemistry in Modern Industrial Boilers with Permission from the American Society of Mechanical Engineers.*

SUGGESTED WATER CHEMISTRY LIMITS
INDUSTRIAL WATERTUBE, HIGH DUTY, PRIMARY FUEL FIRED, DRUM TYPE

Makeup water percentage: Up to 100% of feedwater
Conditions: Includes superheater, turbine drives, or process restriction on steam purity
Steam purity [7]: 1.0 ppm(mg/l) TDS maximum

Drum Operating Pressure	psig 0–300 (MPa) (0–2.07)	301–600 (2.08–4.14)
Feedwater (3)		
Dissolved oxygen ppm (mg/l) O_2 measured before chemical oxygen scavenger addition [1][2]	<0.007	<0007
Total iron ppm (mg/l) Fe	<0.1	<0.05
Total copper ppm (mg/l) Cu	<0.05	<0.025
Total hardness ppm (mg/l) *	<0.5	<0.3
pH @ 25°C	8.3–10.5	8.3–10.5
Nonvolatile TOG ppm (mg/l) C [6]	<1	<1
Oily matter ppm (mg/l)	<1	<1
Boiler Water		
Silica ppm (mg/l) SiO_2	<150	<90
Total alkalinity ppm (mg/l) *	<1000 [5]	<850 [5]
Free OH alkalinity ppm (mg/l) * [4]	NS	NS
Specific conductance mmhos/cm (ms/cm) @ 25°C without neutralization	<7000 [5]	<5500 [5]

* as $CaCO_3$ NS = not specified

(1) Values in the table assume existence of a deaerator.

(2) Chemical deaeration should be provided in all cases, especially if mechanical deaeration is nonexistent or inefficient

(3) Boilers with relatively large furnaces, large steam release space and internal chelant, polymer, and/or antifoam treatment can often tolerate higher levels of feedwater impurities than those in the table and still achieve adequate deposition control and steam purity. Removal of these impurities by external pretreatment is always a more positive solution. Alternatives must be evaluated as to practicality and economics in each individual case. The use of some dispersant and antifoam internal treatment is typical in this type of boiler operation; therefore, it can tolerate higher feedwater hardness than the boilers in Table 1

(4) Minimum and maximum hydroxide alkalinities must be individually specified by a qualified water treatment consultant with regard to silica solubility and other components of internal treatment.

(5) Alkalinity and conductance values are consistent with steam purity limits in the same table. Practical limits above or below tabulated values should be individually established by careful steam purity measurements.

(6) Nonvolatile TOC is that organic carbon not intentionally added as part of the water treatment program.

(7) This limit represents steam purity that should be achievable if other tabulated water quality values are maintained. The limit is not intended to be nor should it be construed to represent a boiler performance guarantee.

Table 5–4 *ASME Water Chemistry Guidelines — 2*
Reproduced from Consensus on Operating Practices for the Control of Feedwater and Boiler Water Chemistry in Modern Industrial Boilers with Permission from the American Society of Mechanical Engineers.

CHAPTER 5: CHEMICAL TREATMENT PROGRAMS
FOR STEAM GENERATING SYSTEMS

SUGGESTED WATER CHEMISTRY LIMITS
INDUSTRIAL FIRETUBE, HIGH DUTY, PRIMARY FUEL FIRED

Makeup water percentage: Up to 100% of feedwater
Conditions: Includes superheater, turbine drives, or process restriction on steam purity
Steam purity [7]: 1.0 ppm(mg/l) TDS maximum

Drum Operating Pressure	psig 0–300 (MPa) (0–2.07)
Feedwater (3)	
Dissolved oxygen ppm (mg/l) O_2 measured before chemical oxygen scavenger addition [1][2]	< 0.007
Total iron ppm (mg/l) Fe	< 0.1
Total copper ppm (mg/l) Cu	< 0.05
Total hardness ppm (mg/l) *	< 1.0
pH @ 25°C	8.3–10.5
Nonvolatile TOG ppm (mg/l) C [6]	< 10
Oily matter ppm (mg/l)	< 1
Boiler Water	
Silica ppm (mg/l) SiO_2	< 150
Total alkalinity ppm (mg/l) *	< 700 [5]
Free OH alkalinity ppm (mg/l) * [4]	NS
Specific conductance μmhos/cm (μs/cm) @ 25°C without neutralization	< 7000 [5]

* as $CaCO_3$ NS = not specified

(1) Values in the table assume existence of a deaeratar.

(2) Chemical deaeration should be provided in all cases, especiahy if mechanical deaeration is nonexistent or inefficient.

(3) Firetube boilers of conservative design, with internal chelant, polymer, and/or anti-foam treatment can often tolerate higher levels of feedwater impurities than those in the table [<0.5 ppm (mg/l) Fe, <0.2 ppm (mg/l) Cu, ≤10 ppm (mg/l) total hardness] and still achieve adequate deposition control and steam purity. Removal of these impurities by external pretreatment is always a more positive solution. Alternatives must be evaluated as to practicality and economics in each individual case.

(4) Minimum and maximum levels of hydroxide alkalinity must he individually specified by a qualified water treatment consultant with regard to silica solubility and other components of internal treatment.

(5) Alkalinity and conductance guidelines are consistent with steam purity target. Practical limits above or below tabulated values should be mdividually established for each case by careful steam purity measurements.

(6) Nonvolatile TOC is that organic carbon not intentionally added as part of the water treatment program.

(7) Target value represents steam purity that should be achievable if other tabulated water quality values are maintained. The target is not intended to be, nor should it be construed to represent, a boiler performance guarantee.

Table 5–5 *ASME Water Chemistry Guidelines — 3*
Reproduced from Consensus on Operating Practices for the Control of Feedwater and Boiler Water Chemistry in Modern Industrial Boilers with Permission from the American Society of Mechanical Engineers.

Boiler Water Treatment Programs

Various treatment programs have been developed over the years to optimize boiler water chemistry control. For the medium- and high-pressure units common to electric utilities and some industrial plants, the most popular treatment programs are coordinated-congruent phosphate, equilibrium phosphate treatment (EPT), phosphate treatment (PT), polymer treatments, caustic treatment, all-volatile treatment, and oxygenated treatment.

Early Boiler Water Treatment

In the early days of steam generation, various odd-sounding treatment methods were used. Some of these treatments had operators putting sawdust or potato peels into boilers. These natural products contributed organics such as lignins, tannins, and starch to the water. The chemicals sequestered hardness ions. Other chemistry programs actually allowed the formation of calcium carbonate on tube walls to protect the tube surface from the boiler water.

These programs were successful because boiler pressures were very low. As boilers increased in size and pressure, more advanced treatment methods became necessary. A big breakthrough came when phosphate treatment was introduced. Sodium phosphate treatments (principally trisodium phosphate (Na_3PO_4) blended with smaller amounts of disodium phosphate (Na_2HPO_4) have been the standard for years in many boiler water conditioning programs. The phosphate ion is effective because it conditions boiler water in an alkaline pH range, and reacts with scale-forming compounds to produce soft sludges. Figure 5–9 shows the corrosion characteristics of carbon steel in boiler water throughout the pH range. Trisodium phosphate provides the needed alkalinity as follows:

$$Na_3PO_4 + H_2O \rightarrow NaOH + Na_2HPO_4 \qquad [Eq.\ 5\text{--}13]$$

Phosphate ions (PO_4^{-3}) can exist in the mono-, di-, and trihydrogen state in aqueous solutions, and thus can give up or accept hydrogen ions. This buffering capability makes phosphate effective in preventing wide pH swings in boiler water due to contaminant ingress.

Phosphate's second major function is to absorb contaminants that enter the boiler. Phosphate reacts directly with calcium to produce calcium hydroxyapetite:

$$10Ca^{+2} + 6PO_4^{-3} + 2OH^- \rightarrow 3Ca_3(PO_4)_2 \cdot Ca(OH)_2\downarrow \quad [Eq.\ 5\text{--}14]$$

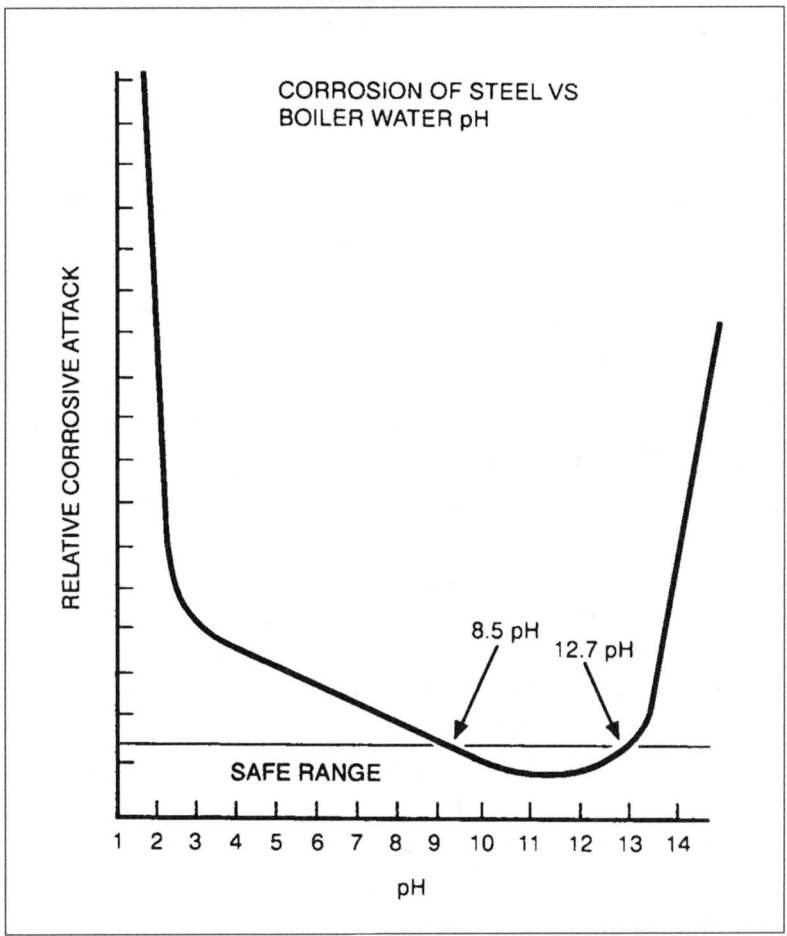

Fig. 5–9 *Corrosion Characteristics of Carbon Steel vs. pH*
Source: Betz Handbook of Industrial Water Conditioning, Ninth Edition. BetzDearborn is a division of Hercules, Inc.

Magnesium and silica react with the alkalinity produced by phosphate to form the nonadherent sludge, serpentine:

$$3Mg^{+2} + 2SiO_3^- + 2OH^- + H_2O \rightarrow 3MgO \cdot 2SiO_2 \cdot 2H_2O\downarrow \qquad [Eq.\ 5\text{--}15]$$

As you may recall, calcium hydroxyapetite and serpentine were two of the boiler tube deposits listed in chapter 4. However, these products exist as soft sludges and are much easier to remove than the hard scale or corrosive products which would otherwise form. They typically settle in the mud drum from which they are removed by blowdown.

Coordinated/Congruent Phosphate Treatment

In the early days of phosphate treatment, phosphate and pH were maintained at fairly high ranges of 20 to 40 ppm phosphate and pH of 11 to 12. As boilers increased in size and pressure, corrosion of the waterwall tubes began to be a problem. Researchers determined that the high alkalinity generated caustic corrosion via the mechanism outlined in Equation 5–13. Accordingly, chemists began to refine phosphate treatment programs to prevent caustic attack. One of the principal programs that came out of this effort was coordinated phosphate treatment.

As is evident from its chemical formula, the molar ratio of sodium to phosphate in trisodium phosphate is three to one. In a coordinated phosphate program, enough disodium phosphate is added to maintain a Na/PO_4 ratio between 2.8 and 2.2 to 1. The Na_2HPO_4 shifts the equilibrium (remember LeChâtelier's Principle) of Equation 5–13 to the left, which helps minimize the formation of free caustic. Figure 5–10 illustrates the phosphate/pH control limits to maintain coordinated treatment at subcritical boiler pressures.

Although coordinated treatment was a significant refinement to phosphate treatment, the process did not stop there. Sodium phosphates are reversely soluble at temperatures above about 250°F, and will begin to precipitate (hide out) on boiler tube walls. The precipitate usually contains a lower sodium-to-phosphate ratio than that in the bulk boiler water, and this can influence the chemistry. As chemists began to recognize this phenomenon, they modified coordinated phosphate treatment to

CHAPTER 5: CHEMICAL TREATMENT PROGRAMS
FOR STEAM GENERATING SYSTEMS

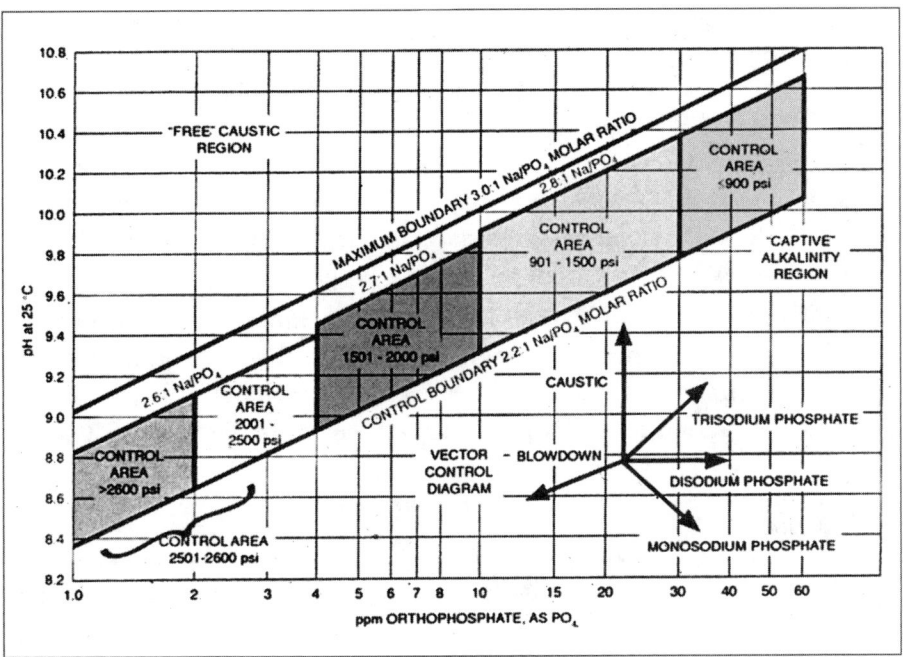

Fig. 5-10 *Coordinated Phosphate Control Diagram*
*Source: Betz Handbook of Industrial Water Conditioning,
Ninth Edition. BetzDearborn is a division of Hercules, Inc.*

congruent treatment, in which the ideal Na/PO_4 ratio became 2.6 to 1 with a lower limit of 2.2 to 1. Evidence at the time indicated that the Na/PO_4 ratio of phosphate deposits was around 2.6-to-1, and this ratio was selected for the treatment so that any phosphate which did come out of solution would precipitate "congruently" and not affect chemistry. That congruent precipitation has been proven to be untrue in many cases, especially in high-pressure units, will be discussed in the next section.

Congruent treatment and occasionally coordinated treatment are still widely used in many low- and medium-pressure boilers, and in high-pressure utility boilers that do not suffer from hideout. Phosphate concentrations in a congruent program are typically maintained within a 2 to 5 ppm range. The key to the program is the sodium-to-phosphate ratio. If the ratio climbs above the recommended limit of 2.6:1, the water may become too alkaline. Conversely, ratios below the recommended lower limit of 2.2:1 can generate acidic conditions.

In a water solution containing only the phosphate species, or phosphate plus non-reactive compounds, the sodium-to-phosphate ratio can be calculated directly from the phosphate concentration and pH. In fact, more than one computer program has been published to perform these calculations. However, certain chemicals, most notably ammonia or amines in steam-generating systems, will influence pH and render standard calculations unusable. Typically, ammonia or amines raise the pH, which gives artificially high sodium-to-phosphate ratios.

This effect must be taken into account when monitoring a coordinated or congruent phosphate program. Several methods are possible. Figure 5–11 illustrates control curves for coordinated phosphate treatments at various ammonia levels. By monitoring the ammonia concentration in the

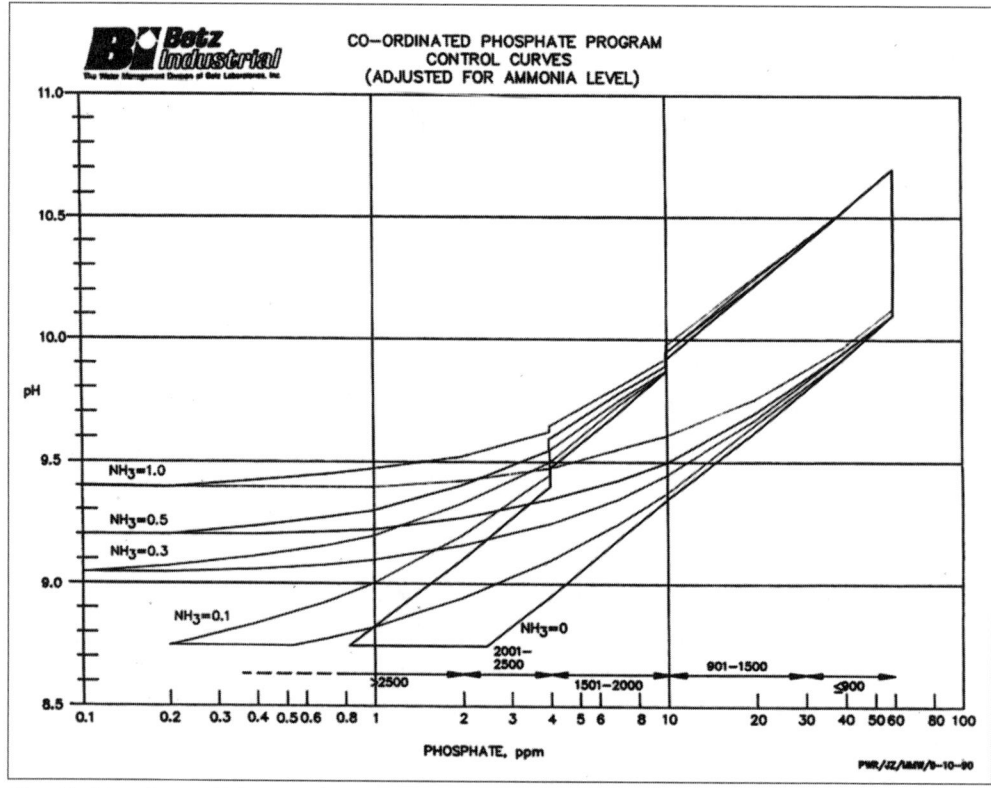

Fig. 5–11 *Control Curves for Coordinated Phosphate Treatments at Various Ammonia Concentrations*
Source: Betz Handbook of Industrial Water Conditioning, Ninth Edition. BetzDearborn is a division of Hercules, Inc.

boiler water, the chemist can prepare phosphate dosages based on the curves. Where neutralizing amines are used, the calculation becomes more complicated since various amines exhibit different basicities. At least one firm has developed a computer program that accounts for the effects of ammonia and the common amines that are used in feedwater systems. Personally, I would consider continuous on-line analyses of both phosphate and sodium of the boiler water. These would allow direct calculation of the sodium-to-phosphate ratio. If done on-line, the results can be incorporated into data acquisition systems.

Within the last decade or so, some difficulties have arisen with coordinated/congruent chemistry in high-pressure boilers. This is due to phosphate hideout and the incongruent precipitation of sodium phosphate.

Phosphate Hideout

We have already briefly mentioned the reverse solubility of sodium phosphate. Trisodium phosphate is most soluble at 250°F, but then solubility rapidly falls off and becomes very low at 600°F. The effect may become very pronounced in units that operate above 2,000 psi, and/or are subjected to frequent load changes. When phosphate comes out of solution, the solids precipitate on tube walls and other boiler internals. This phenomenon is known as hideout. Reports by some utilities formerly using congruent treatment indicate that phosphate concentrations would often drop well below 1 ppm when hideout was severe. This depleted the boiler water of the chemical designed to control chemistry. Researchers also found that the phosphate compounds often precipitated incongruently, with deposit sodium-to-phosphate molar ratios of 2 to 1 or lower.

Hideout can be at its worst in a cycling unit. Consider a boiler at low load with congruent treatment. As load is raised and heat fluxes increase, sodium phosphate begins to precipitate onto the tube surfaces. Due to incongruent precipitation, the Na/PO_4 ratio of the phosphate remaining in solution rises. This increases the pH to perhaps several tenths of a unit above congruent guidelines. The chemist may then add a di-/trisodium blend to raise the phosphate concentration and lower the pH, but this phosphate also hides out. When boiler load is reduced, the situation reverses and the precipitated phosphate redissolves. In this case, however, the sodium phosphates, which have a low Na/PO$_4$ ratio, drive the pH downward. Severe hideout and the reverse dissolution process have been known to force boiler water pH below 7 in cycling units. Since EPRI recommends unit shutdown if the pH drops below 8.0, such events can be disturbing. Coordinated or congruent phosphate treatment may be impossible to control in boilers that exhibit this phenomenon.

Phosphate hideout appears to be further influenced by the cleanliness of the boiler tubes, and becomes more severe with increased deposit loading, particularly iron oxides. The evidence indicates that sodium phosphates form a sodium-iron-phosphate complex with the magnetite layer. Routine boiler chemical cleanings can potentially reduce hideout by minimizing the formation of porous magnetite deposits.

The sodium-iron-phosphate reaction may cause other problems. Several reports have been published which suggest that sodium phosphate actually participates in corrosion reactions. This may in part be due to the fact that the phosphate hides out at a low sodium-to-phosphate ratio, which tends to make the deposits acidic. Acid phosphate corrosion is now suspected as being a significant culprit in many boiler tube failures that previously were thought to have been caused by other mechanisms.

At this point a partial rebuttal is in order. Not all boiler water experts are convinced that coordinated/congruent phosphate treatment deserves the reputation it seems to have acquired, except perhaps for very high-pressure boilers. One well-known expert still thinks that the programs are suitable on boilers of up to 2800 psig in pressure. Personally, I prefer a case-by-case evaluation. My former utility operates a 2400 psig boiler. This unit is equipped with a continuous on-line phosphate monitor. Hideout has never been a major problem.

Despite continued uncertainty regarding coordinated and congruent phosphates, many utilities have switched to alternative programs. The next section describes two alternative phosphate programs. Following that is a discussion of non-phosphate programs, some of which have been around for a long time and some of which are a bit more modern.

Chapter 5: Chemical Treatment Programs for Steam Generating Systems

Alternative Phosphate Treatment Programs

At least two alternative phosphate programs have been developed to counteract the effects of hideout or acidic phosphate corrosion. These are equilibrium phosphate treatment (EPT) and phosphate treatment (PT).

Equilibrium Phosphate. EPT was pioneered by Jan Stodola of Ontario Hydro for use in the utility's high-pressure fossil units. In an EPT program, phosphate residuals are maintained within a range of 0.2 to 2.5 ppm, and pH within a range of 9.0 to 9.7. Alkalinity is controlled by the addition of trisodium phosphate with supplemental addition of caustic, generally at unit startup. This treatment produces a solution with a sodium-to-phosphate ratio of 2.8:1 or greater, with most of the control range at 3:1 or above. Caustic alkalinity is maintained from 0 to 1 ppm.

While equilibrium phosphate programs minimize hideout, the treatment is less forgiving towards contaminant in-leakage than coordinated or congruent programs. Thus, monitoring of contaminant ingress, as outlined in chapter 6, is very important. Another potential problem is that the higher pH and higher caustic levels can lead to an increase in carryover of sodium hydroxide to the turbine. NaOH is a contributor to stress-corrosion-cracking (SCC) in turbine blades.

Phosphate Treatment. Phosphate treatment almost appears to be a combination of EPT and congruent/coordinated treatment. Phosphate residuals are maintained between 2.5 to near 10 ppm, with a sodium-to-phosphate ratio of 2.8:1 or greater. As with EPT, caustic alkalinity is held to 1 ppm maximum. Hideout in PT-conditioned systems has been reported. Although hideout deposits are more alkaline than those in a congruent program, the hideout process can still create difficulties for the chemist. Possibly for this reason, EPT appears to be the more preferred phosphate-based replacement for congruent or coordinated programs.

Other Boiler Water Treatment Programs. Phosphate programs are not the only methods available for boiler water treatment. In some cases, they are totally unsatisfactory. For example, once-through boilers cannot tolerate dissolved solids, and this prohibits phosphate use.

Chelants and Polymer Treatment. Chelants are compounds that chemically bind (chelate) certain elements. Two of the most commonly-used chelants for boiler water treatment are ethylenediamine-tetraacetic acid or EDTA (Figure 5–12), and nitrilotriacetic acid. These chelants complex cations such as calcium, magnesium, iron, and copper through interaction of the positively charged ions with the partially negative oxygen and nitrogen atoms of the chelant. The cation in a chelated structure

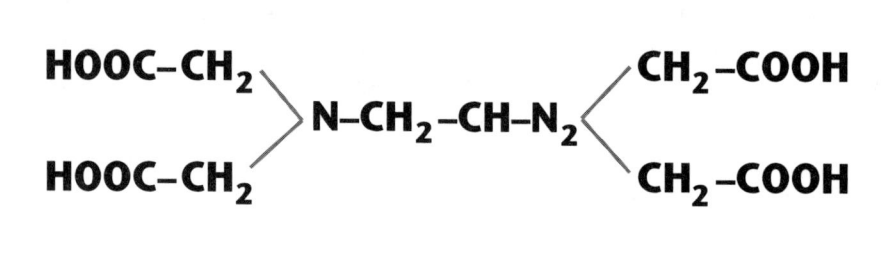

Fig. 5–12 *Structure of EDTA*

appears as if it is in a cage. Chelants do not form precipitates like the phosphates, but rather keep hardness and other ions in solution so that they can be removed in the boiler blowdown. EDTA has been effective in boilers operating at pressures up to 1,500 psig, but the compound will begin to break down at a temperature of 400°F to produce byproducts that also chelate, but less effectively. NTA is typically limited to boilers of less than 900 psig pressure.

The chelants react on a 1:1 basis with cations, thus a significant residual may be needed to tie up contaminants. An EDTA molecule has a molecular weight of 288, while the respective molecular weights of calcium and magnesium are 40 and 24.3. This indicates that 7.2 ppm and 11.9 ppm of EDTA would theoretically react with 1 ppm of calcium and magnesium, respectively.

The use of chelants must be carefully considered, because they can cause problems in boiler systems. For example, chelants are attacked by dissolved oxygen, and should not be added to the feedwater until after the deaerator. More importantly, improper control and overfeed of chelants can cause dissolution of the protective magnetite layer on the boiler tubes. Serious corrosion has been known to occur.

Polymers of the acrylate, sulfonated styrene, and other families are also used for boiler water scale control. Most often the polymers are employed as part of a combined program with phosphates or chelants. The polymers modify the crystalline structure of precipitates, making them less sticky and more easily removable by blowdown. These chemicals act on a sub-stoichiometric basis with the compounds being treated, and typical dosages may range from 1 to 10 ppm.

One area where sequesterants such as polymethacrylate or sulfonated styrene appear to be particularly effective is iron control, especially after unit startup. The sequesterants will bind free iron and allow it to be removed by the boiler blowdown rather than precipitate on the tube walls.

Chelants and polymers have been effectively used for boiler water chemistry control. But, proper selection of the correct chemical or combination of chemicals may not be easy. Monitoring of the chemical residuals is more difficult than in phosphate programs. Many people in the electric utility industry appear to favor the more "simple" phosphate or volatile treatment programs. For low-pressure industrial treatment, however, a chelant/polymer program may perform very well. Certainly polymers should be considered for iron control during unit startups.

Caustic Treatment. Caustic treatment, wherein sodium hydroxide is the principal boiler water conditioning chemical, is used at a number of overseas utilities. In a caustic treatment program, boiler water pH is maintained within a range of 9.4 to 9.6 by addition of caustic to maintain a NaOH level of 1.0 to 1.5 ppm. Feedwater chemistry in a caustic treatment program must be well controlled to prevent excessive deposit formation in the boiler, which might lead to under-deposit caustic corrosion. Caustic treatment of high-pressure boilers has not caught on in the United States, although caustic control of low-pressure boiler water chemistry was once a common technique. Sodium nitrate is frequently added with the caustic to prevent stress corrosion cracking of boiler tubes. As is the case with equilibrium phosphate treatment, the higher levels of caustic in the boiler water may lead to greater carryover of sodium hydroxide to the turbine.

All-Volatile Treatment. All-volatile treatment (AVT) was principally developed for once-through boilers. Like oxygenated treatment, AVT is another of the combined feedwater/boiler water treatment programs. The primary treatment chemicals, ammonia or amines and an oxygen scavenger carry through the boiler into the steam. AVT guidelines for a once-through unit call for a pH range of 9.3 to 9.6 with less than 2 ppm dissolved solids. Ammonia levels may range from 1 ppm to 2 ppm. Condensate polishers are again an absolute requirement.

AVT is also used in some very high-pressure, drum-type units. As pressure approaches the critical point of 3204 psig, the density of steam and water become nearly equal, and at the critical point, water and steam turn into a single phase. Even in sub-critical boilers operating as low as 2800 psig, it is difficult to separate water from steam in the boiler drum internals, so boiler water must be quite pure to prevent carryover. The advantage of AVT is that it does not introduce dissolved solids to the feedwater, which subsequently might carry over. But, AVT does not protect drum boilers from contaminant introduction due to a condenser leak or other problem. Condensate polishers are the most effective buffer against chemistry upsets, and are a virtual requirement for reliable operation of any AVT unit, even drum boilers. I did actually visit one utility that switched their 2400-psig drum unit from congruent phosphate to AVT, with no intention of installing a polisher. So far, they have not had any major upsets, but I would be concerned. In a case like this, the chemical feed arrangement should contain an emergency phosphate feed system so that phosphate can be immediately injected into the drum in the event of an upset. This must be considered as only a short-term, temporary measure, because the sudden increase in dissolved solids due to contaminant in-leakage will greatly increase the potential for carryover and deposition.

Several problems with AVT have become evident besides its sometimes inefficient performance in feedwater systems. First, the volatile treatment chemicals tend to increase carryover of chloride and sulfate, which then deposit on the low-pressure turbine blades. Second, the volatiles are less capable of neutralizing acidic turbine deposits than phosphate, which, if it carries over in small quantities, may be beneficial due to its alkaline nature. Third, amines and organic oxygen scavengers can break down into carbon dioxide and organic acids. These are potential corrodents of turbine blades and afterboiler components.

CHAPTER 5: CHEMICAL TREATMENT PROGRAMS
FOR STEAM GENERATING SYSTEMS

AVT has lost a lot of its popularity. Oxygenated treatment has supplanted it at many once-through utilities overseas and now in the United States.

Oxygenated Treatment. We have already examined OT as a feedwater treatment, and have seen that its effectiveness in boilers stems from the fact that OT greatly reduces iron transport. European utilities that have been using OT for a long time have found that chemical cleanings have been significantly reduced. Practical Example 5–1 provides some specific details on the startup and initial observations of an OT program at a supercritical steam generating unit.

Practical Example 5-1

In this 400 MW, 3500 psig supercritical unit, plant personnel switched from AVT to OT, in part because all-volatile-treatment required a boiler chemical cleaning every 18 months to 2 years. They selected bottled gaseous oxygen as the treatment chemical, and installed injection quills at the condensate polisher outlet and the deaerator outlet. An oxygen flow rate of 2.2 SCFH maintains a 50 to 150 ppb concentration through all load conditions. Supplemental ammonia feed keeps the condensate/feedwater pH within a range of 8.0 to 8.5.

The plant staff observed the following conditions after the program was started.

- The condensate portion of the piping developed the protective FeOOH film in about one day. It took almost a week for the feedwater piping to develop the layer.

- The cation conductivity of the condensate remained below the recommended 0.15 US guideline.

- Dissolved iron concentrations dropped from a typical norm of about 9 ppb to around 3 ppb.

- The black magnetite surface became brownish-orange due to the formation of FeOOH.

Heat Recovery Steam Generators

Boiler water treatment for HRSGs may take on some unique aspects, depending on the unit configuration. As we learned earlier, one common HRSG design uses the low-pressure boiler as the feedwater source for an intermediate and high-pressure circuit. Such a configuration requires high-purity makeup, zero-solids treatment for the low-pressure circuit, and perhaps even condensate polishing of the feedwater. Treatment programs for the remainder of the HRSG can be quite varied. One combined-cycle manufacturer fabricates HRSGs with a low-pressure circulating stage and a high-pressure, once-through stage. The manufacturer recommends AVT for the low-pressure circuit and OT for the high-pressure system. Phosphate treatments can be effective in some HRSGs, because the lower heat fluxes minimize hideout. A very important factor is whether the unit is base loaded or cycled. If the latter, then chemistry control will be difficult regardless of the program.

A very important item of concern for HRSGs is flow-assisted-corrosion. FAC, as we have seen occurs in conventional steam generating systems, where water velocities typically range from 1.5 to 12 feet per second. The velocities in HRSGs may easily exceed 15 fps, which can generate FAC in the short-radius elbows that are common with these units. A couple of solutions are possible. One is oxygenated treatment of the unit. However, this severely limits treatment program flexibility. Another solution is to design the unit with elbows fabricated from a higher-grade of steel such as the $2\frac{1}{4}$–$\frac{1}{2}$ chrome-moly alloy that is used for some boiler tubes. This material is resistant to FAC. A well-designed HRSG gives the operator flexibility in selection of water treatment methods.

Steam Chemistry

Recommended steam purity guidelines are shown in Tables 5–6 and 5–7. The latter table illustrates figures developed a while back by two of the major turbine manufacturers. These are the values that the manufacturers believe will safeguard their turbines from damage due to deposition and corrosion by contaminants. As we have seen, most steam turbine chemistry problems are caused by injection of contaminated feedwater

Steam Purity Guidelines by Treatment

Parameter	Equilibrium Phosphate Treatment All Volatile Treatment (AVT) Oxygentated Treatment	Phosphate Treatment
Sodium (ppb)	3	5
Cation Conductivity (μs/cm)	<0.15	<0.3
Chloride (ppb)	3	3
Sulfate (ppb)	3	3
Silica (ppb)	10	10
TOC (ppb)	100	100

Table 5-6 *Recommended Steam Purity Guidelines — 1*
Source: EPRI

Steam Purity Guidelines by Turbine

Parameter	Vendor #1	Vendor #2
Sodium (ppb)	5	<3
Cation Conductivity (μs/cm)	<0.3	<0.2
Chloride (ppb)	<5	
Silica (ppb)	<10	
Copper (ppb)	<2	
Iron (ppb)	<20	

Table 5-7 *Previously Published Steam Purity Guidelines — 2*

through the attemperator system or carryover of contaminants from the boiler. The former can be prevented by good control of feedwater chemistry. However, the latter scenario may be the result of any number of factors. These include poor drum design, fast load swings that cause sudden jumps in drum level, introduction of contaminants to the boiler water that cause foaming, and poor water chemistry control that allows excessive amounts of impurities to concentrate in the boiler water. The first two of these are the responsibility of the boiler design firm and boiler operators, respectively. The last two are chemical in nature.

We have already looked at foaming, so let's consider boiler water chemistry control. Most boiler water impurities are soluble to some extent in steam, although a few, most notably silica, will vaporize and carry over on their own. Solubility and vaporous carryover both increase with increasing pressure. Therefore, higher boiler pressures require better quality feedwater. Table 5–8 shows general guidelines for four of the most problematic turbine contaminants, and Figures 5–13 and 5–14 show the change in silica volatility versus pressure. The allowable concentrations drastically decrease with increasing pressure. Please note that these are general guidelines and that each steam-generating unit should be monitored individually. This is one of the reasons why I have devoted a whole chapter, the next one, to sampling. Boiler water guidelines that may be fine for one unit could be quite inadequate for another. Only by obtain-

Recommended Boiler Water Concentrations to Meet Steam Purity Guidelines

Boiler Pressure (psig)	Sodium (ppm)	Silica (ppm)	Chloride (ppm)	Sulfate (ppm)
900	3.3	3.6	0.33	0.33
1100	3.0	1.9	0.24	0.24
1300	2.7	1.3	0.17	0.17
1500	2.5	0.81	0.13	0.13
1700	2.2	0.57	0.086	0.086
1900	1.9	0.39	0.065	0.065
2100	1.6	0.27	0.048	0.048
2300	1.2	0.19	0.037	0.037
2500	0.71	0.14	0.028	0.028
2700	0.44	0.085	0.020	0.020
2900	0.27	0.051	0.014	0.014

Table 5–8 *General Guidelines for Turbine Contaminants*
Source: EPRI

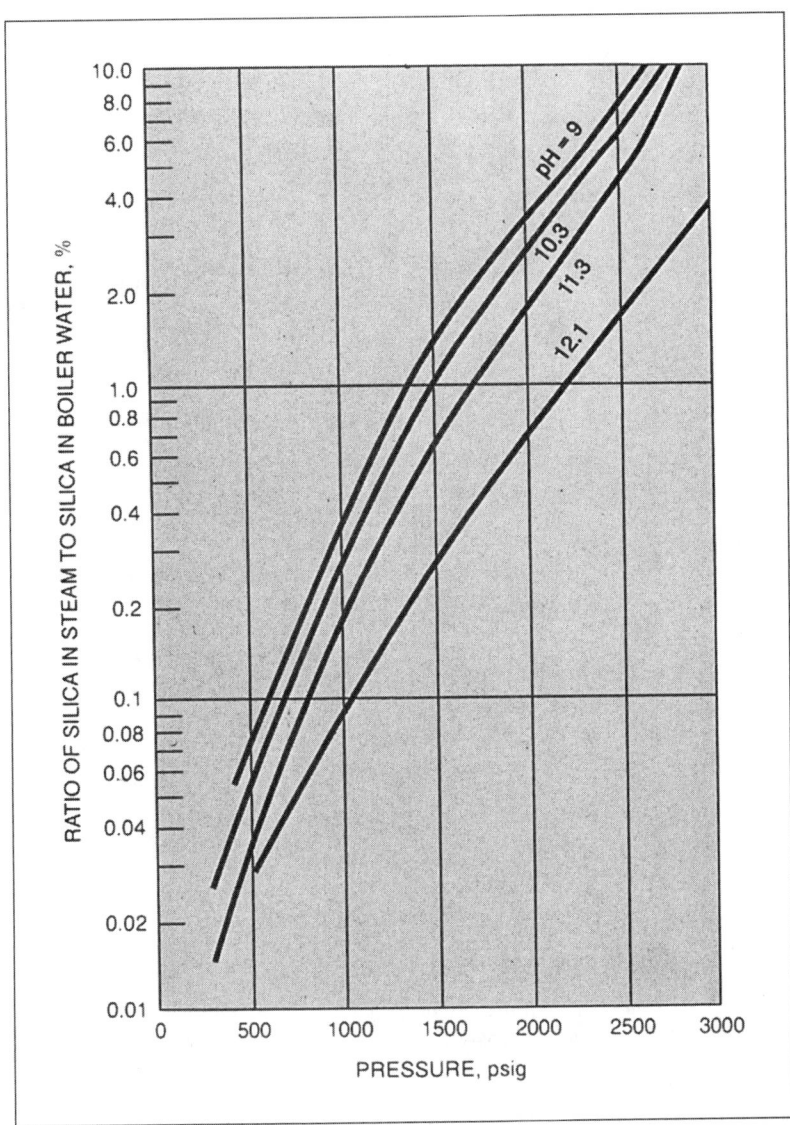

Fig. 5–13 *Change in Silica Volatility vs. Pressure*
Source: Betz Handbook of Industrial Water Conditioning, Ninth Edition. BetzDearborn is a division of Hercules, Inc.

ing representative data can the plant chemist, operator, or engineer make an intelligent decision about chemistry factors.

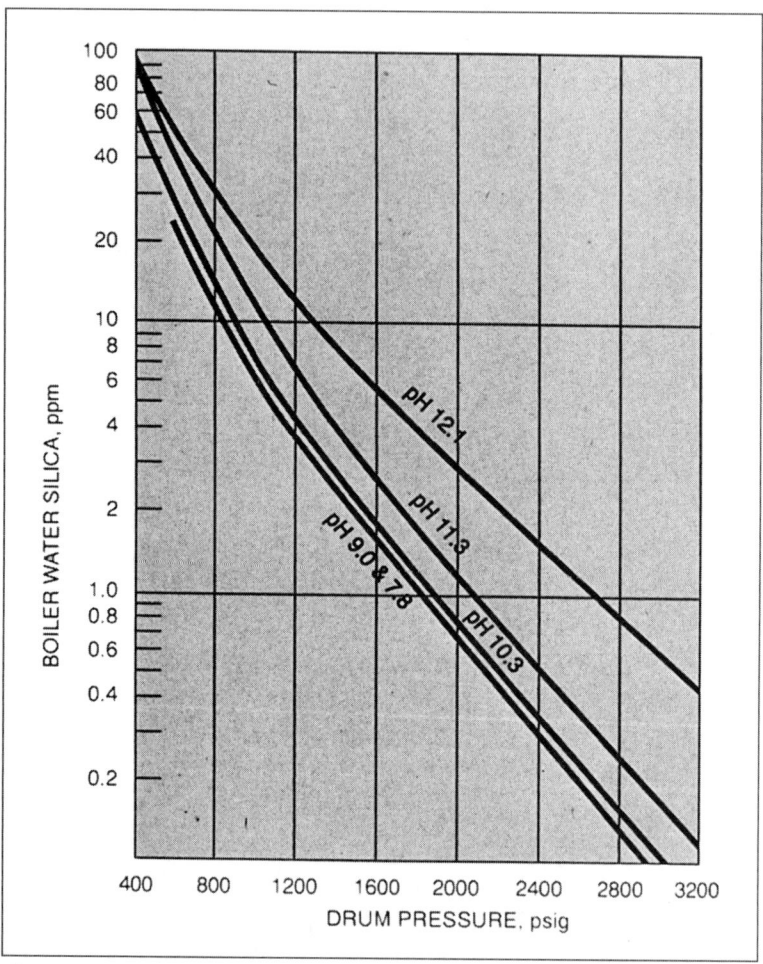

Fig. 5–14 *Allowable Drum Silica Concentrations vs. Pressure*
Source: Betz Handbook of Industrial Water Conditioning, Ninth Edition. BetzDearborn is a division of Hercules, Inc.

Condensate Return

Chemical control of condensate return from an industrial process is also important. If contaminants are introduced from the industrial process, then perhaps condensate polishing is the only method to treat the water so it can be returned to the boiler. Corrosion prevention may be another matter. Carbonic acid and oxygen corrosion are the two most common mechanisms. Several control methods may be possible. One is selection of an amine with a good steam-to-water distribution ratio so that the amine carries over into the condensate system. The dosage of amine would have to correlate with the calculated amount of CO_2 that carries over also. Another possibility is injection of the neutralizing amine directly into the condensate system. Then, boiler water chemistry could be maintained independently. A third alternative is to inject filming amines into the condensate system. Filming amines are long-chain organic molecules (typically 10 to 20 carbon atoms) that have the amine group on one end. These function differently than neutralizing amines. The amine ends of the molecules actually attach to the metal surface and then the organic ends form a hydrophobic (water-hating) layer that protects the pipe wall. Dosages of only a few ppm are required, although the flowing stream has a tendency to continually wash away the film. Also, inadvertent overfeed of filming amines can send sticky masses downstream to foul equipment such as condensate polishers. They must be used with caution.

Layup and Off-Line Corrosion Protection

More corrosion can occur during an outage than at any other time. This is almost exclusively due to air intrusion into the boiler and subsequent oxygen corrosion of components. Carbon dioxide that enters with air will form carbonic acid and cause additional corrosion. The optimum theoretical approach to prevent corrosion is to completely dry the boiler with warm air circulation. For long-term outages, where the unit may not be needed for quite a while, this merits consideration, although complete dryness is sometimes difficult to achieve. Silica-gel desiccants, placed at strategic locations in the boiler will help.

Many outages are only for a few days or a few weeks, with the knowledge that the boiler will be required to operate on a specific date. Wet layups are more practical for these situations. But, the boiler and feedwater system must be laid up properly or severe corrosion may result.

Layup guidelines differ depending on whether the layup will be short-term, less than four days or so, or long term, such as several weeks or a month. One short-term layup method calls for the boiler to be filled with condensate containing 200 ppm of hydrazine and enough ammonia or neutralizing amine to bring the pH to 10. A nitrogen blanket should be introduced to the drum and superheater through the vent lines as the boiler pressure decays. The feedwater system must also not be neglected. If the standard dosages of oxygen scavenger and ammonia or amines remained in the feedwater at time of shutdown, this solution can remain. If not, then provisions should be made to introduce a solution containing 50 ppb of oxygen scavenger at a pH of 10 to the system. The steam side of the heaters should be blanketed with steam or nitrogen, as exfoliation of feedwater heater tubes can become very severe without this protection. The deaearator should also be blanketed with steam, if possible. The superheater and reheater are usually allowed to remain dry.

For long term outages, some modifications to the treatment are recommended. For example, a solution containing 50 to 100 ppm of hydrazine is suggested for the feedwater system, with pH adjustment to 9.5. Some experts also recommend filling the superheater with the same solution as in the boiler. If this is done, plant personnel must pin the superheater hangers, as the water will add a great deal of weight to the superheater.

Where possible, the boiler layup solution should be periodically circulated if possible. This helps reduce stagnant zones and also allows the plant chemists to obtain more accurate analyses of the layup chemical concentrations. Should the concentrations be too low, the staff can add more chemicals.

A couple of problems are evident with these procedures. One, the boiler must be drained before light off due to the high concentration of ammonia and oxygen scavenger. The second is the requirement for pinning superheater hangers if it is filled with treated water. Third, circulation of the chemicals may not be easy, especially if the boiler cannot be fired. At the 1998 International Water Conference, an author presented a paper outlining a modified layup procedure. Major steps included:

- Place the boiler and feedwater heaters under a nitrogen blanket. Begin feeding nitrogen just as the unit is coming off-line so that the cooling process will draw nitrogen in, not air. Maintain 1 psig pressure on these vessels. Also place a nitrogen cap on the deaerator.

Chapter 5: Chemical Treatment Programs for Steam Generating Systems

- Feed nitrogen into the condenser but do not pressurize the vessel.
- If necessary, as the unit is coming down increase the oxygen-scavenger feed to the feedwater to obtain a 50 ppb residual.

Where multiple units are together, steam blanketing can replace nitrogen blanketing in the boiler, feedwater heaters, and deaerator. This not only prevents air from entering, but keeps the boiler warm. This eases startup and reduces mechanical stresses.

Chemical Feed Systems

It is very important that chemical feed systems be designed to provide steady dosages to steam-generating systems. Figure 5–15 shows the outline of a possible system. Rather than go into a long-winded discussion about this system, I will point out some of the highlights.

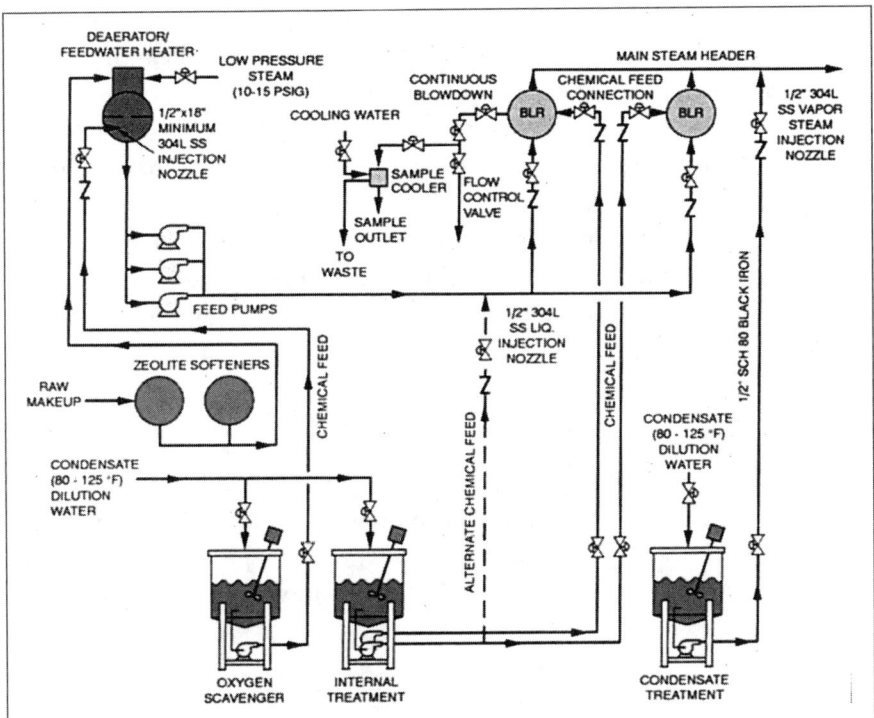

Fig. 5–15 *Chemical Feed System*
Source: Betz Handbook of Industrial Water Conditioning, Ninth Edition. BetzDearborn is a division of Hercules, Inc.

- Feed lines to the steam-generating portion of the system are fabricated from stainless steel. This minimizes corrosion an contaminant introduction through the sample lines.
- Chemical is injected into pipes through injection quills. This provides for better and quicker distribution of the chemical into the process stream.
- Condensate treatment chemicals enter the main steam header.
- Each feed line is equipped with a check valve to prevent back feed of hot water or steam. This is an important safety feature.

An important criteria in setting up a chemical feed system is to size the pump such that it will produce a pressure higher than the maximum system pressure at the point where the chemical is injected. Head losses in the line will reduce the pressure a bit between the pump and injection point.

With the great development of computers and control systems, some chemical feed can be automated. In the on-line water chemistry monitoring system that I helped set up at my former utility, I programmed the PLC to adjust oxygen-scavenger feed rates based on the oxygen-scavenger concentration in the feedwater. Given the present concern about flow-assisted-corrosion, the control could be modified to read dissolved oxygen or ORP data. Many of the major water treatment companies have developed tracer techniques, whereby the treatment systems will monitor and adjust feed based on a particular identifying element added to the treatment chemical.

Conclusion

This chapter outlined chemistry programs that are suitable for feedwater and boiler water treatment. Also presented were control guidelines. These guidelines are impossible to follow without proper sampling and analytical equipment. This is presented in the next chapter.

MONITORING TECHNIQUES AND CONTROL GUIDELINES 6

The preceding chapters have shown how even small amounts of contamination can dramatically influence boiler operation. Proper monitoring and prompt detection of chemistry upsets, especially in higher-pressure systems, are extremely important. The development of reliable and affordable on-line monitoring equipment has made this a more straightforward task. The following chapter is designed to give the reader a practical guide toward sample point selection, sampling parameters, and techniques required to collect representative samples. The discussion is informative for those wishing to set up or improve a sampling and monitoring system. The chapter also includes additional water chemistry guidelines, so that the reader can evaluate performance at his or her plant.

Sample Point Selection

Figure 6–1 is a duplicate of Figure 4–1, but with sample points added. As the reader will observe, the sample points are placed strategically throughout the system, so that all critical processes may be monitored. These same sample points are also important in plants with heat recovery steam generators, and in fact even more monitoring may be required where the HRSG has more than one boiler circuit. The most important sample locations include the following:

- Makeup treatment system
- Condensate return from the industrial plant
- Condensate pump discharge
- Condensate polisher effluent
- Dearator inlet
- Deaerator outlet
- Feedwater or economizer inlet
- Boiler water
- Saturated steam
- Main steam
- Reheat steam

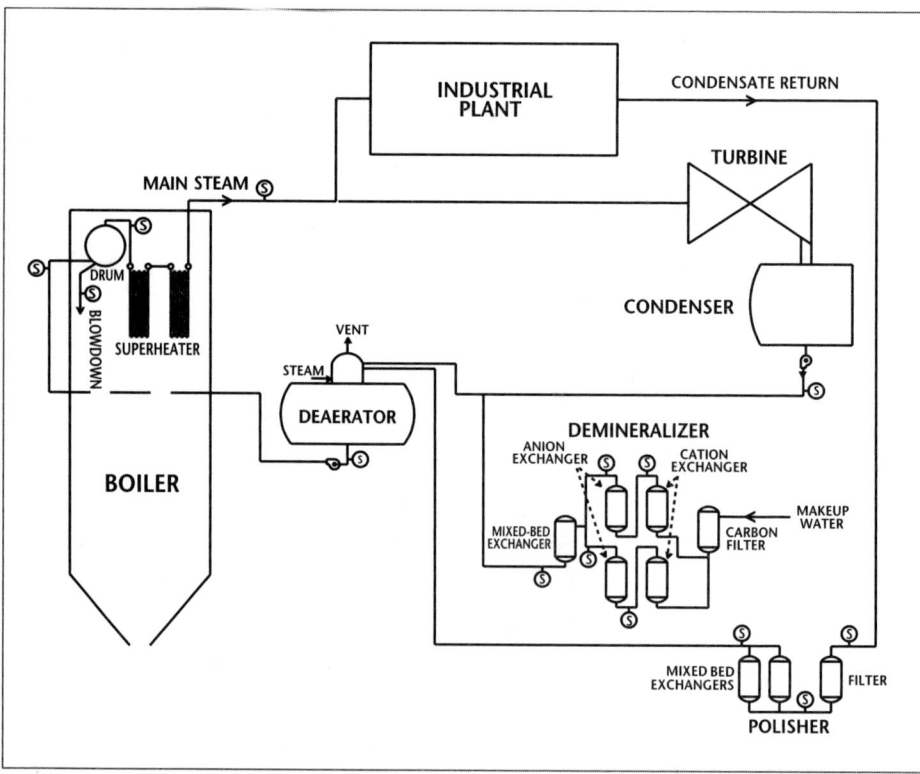

Fig. 6-1 *Recommended Sampling Points in a Steam Generation System*

The following sections discuss each of these sample points in greater detail, and outline recommended analyses. I have used three different categories for sample importance; primary on-line, for those samples where on-line analyses are critical; supplementary on-line, for data to support the primary measurements; and grab sample analyses, which help plant personnel fine tune chemistry or check for constituents that are not readily analyzable on-line.* This chapter is based upon principles found in the literature, but a few of the sampling recommendations also come from my own experience as a utility engineer and chemist. Other experts may suggest additional measurements.

Makeup Treatment System

We have already covered makeup treatment rather extensively in chapter 3, and have seen that the type and complexity of a makeup treatment system depend on the requirements of the steam-generating system. For medium- and high-pressure boilers, and especially those that power turbines very pure makeup is essential. The makeup treatment system shown in Figure 6–1 contains a cation/anion/mixed-bed ion exchange arrangement. Important analyses for this system include these:

Primary On-Line:

- Cation Exchanger Effluent — Specific conductivity, sodium.*
- Anion Exchanger Effluent — Specific conductivity, silica.*
- Mixed-Bed Exchanger Effluent — Specific conductivity,* sodium,* silica,* pH.

Grab

- Mixed-Bed Effluent — Chloride, sulfate, total organic carbon (TOC).[1]

[1] *Some experts recommend on-line sampling for these parameters.*

This list is understandable in light of the principles outlined in chapter 3.

---

* *You will notice an asterisk besides some of the on-line analyses. These indicate measurements for which grab sample analyses can be quite helpful to verify on-line instrument reliability and water quality. During my days as a utility chemist, I analyzed many of these samples using a variety of benchtop instruments including a UV-Vis spectrometer, pH and conductivity meters, and atomic absorption spectrophotometer, and an ion chromatograph. Other methods are possible for some of these analyses, but the important point is that periodic or daily grab sampling helps provide good quality control of steam generation chemistry.*

Sodium is the most weakly held ion on a cation exchange resin, and is the first constituent to come off the bed when the resin reaches exhaustion. Thus, on-line sodium monitoring is a good tool for preventing overrun of the exchanger. Specific conductivity monitoring of the cation effluent by itself is not efficient in determining bed exhaustion; however, comparison of the effluent conductivity with the conductivity of a sample extracted several inches from the bottom of the bed can be used to determine the approach of bed exhaustion. This is known as anticipatory sampling. The technique has not been universally adopted.

Silica is the most weakly held constituent on the anion resin, so anion effluent silica analyses are very helpful for monitoring bed performance and exhaustion. Specific conductivity measurement of the anion exchanger effluent can also provide an effective indication of either anion or cation resin exhaustion. If the anion exchanger exhausts first, the conductivity will slightly but noticeably dip before rising. This is due to the release of silica, which, when combined with the small amount of sodium ions that always leak from a cation bed, forms sodium silicate. This compound is less conductive than the sodium hydroxide that normally elutes. If the cation exchanger exhausts first, the conductivity of the anion effluent will rise without any dip. This is caused by the excess sodium ions escaping from the cation bed, which cause an increase of sodium hydroxide in the anion effluent.

The mixed-bed exchanger polishes the cation/anion effluent and reduces dissolved ion concentrations to part-per-billion (ppb) levels. Because the mixed-bed is the last line of defense before the steam generating system, continuous monitoring is vital to prevent the accidental release of contaminants into the condensate. On-line sodium, silica, and specific conductivity analyses provide a safeguard against bed exhaustion and overrun, and also indicate if the exchanger is performing properly. I included pH in the list of primary on-line instruments, more out of tradition than anything else. Measurement of pH in very pure water is still somewhat problematic, and actually the other recommended analyses will provide a quick indication of an upset. Some experts also recommend analyses for chloride, sulfate, and TOC to provide additional protection to the boiler and steam system.

Other makeup treatment methods are now available for producing high-purity water, and these may require different or supplemental analyses. Consider a reverse osmosis/mixed-bed arrangement, where the RO unit takes the place of the cation/anion demineralizer. Because RO is a filtration process, where pressure forces water through membranes, on-line analyses are more mechanically oriented and usually consist of pressure, flow, temperature, and conductivity. The mixed bed analyses remain the same.

The newer generation of HRSGs are being designed to operate at pressures near or above 2000 psig. These units require good quality make-up produced by equipment with similar instrumentation to that listed above. For lower-pressure boilers of under 600 psig, treatment systems often are much less sophisticated since these boilers are more tolerant to dissolved solids, with the exception of hardness. Sodium softening to remove hardness is a common treatment for low-pressure boilers, so sampling of the softener effluent for calcium and magnesium is important. If the softener is reliable and throughput predictable, then grab sampling may be satisfactory. Better, however, is on-line hardness monitoring.

Condensate Return. We have seen that condensate from industrial processes may pick up contaminants that are either introduced from the plant process(es) or are generated by corrosion in the condensate return lines. Contaminants can include inorganic ions, metal oxide particulates, oils, and other organics.

The recommended analyses of the condensate return vary depending upon the sources and nature of contamination. If organics are present, then TOC is a suitable choice. Specific conductivity is a very popular measurement. At some facilities, the electronic output from the conductivity monitor operates an automatic dump valve, which opens if the monitor detects excess contamination in the condensate return. I have also seen this done with on-line TOC monitors at a petrochemical plant. Obviously, dumping condensate can be expensive if done frequently or for long periods. But, an automatic dumping system does provide a safeguard to the unit.

Recommended condensate return analyses may include these:

Primary On-Line

- pH*
- Specific conductivity*

Supplementary On-Line

- Total Organic Carbon[1]

Grab

- Iron (see discussion below)

[1] *On-line analyzer is recommended if organics are a contaminant.*

Grab sample iron analyses provide valuable data on corrosion in the condensate return line and performance of any neutralizing or filming amines that may be used for corrosion control.

Condensate Pump Discharge. The condensate pump discharge is of course a critical monitoring point, as condensers are often the most likely source for major contamination, especially at electric utilities. Where the condensate is treated in a polisher, the effects of a condenser tube leak are mitigated. However, it is still important that any leak be detected as quickly as possible to prevent premature exhaustion of the polisher and subsequent carryover of contaminants to the boiler. Condenser tube leaks on unpolished systems have been known to cause boiler tube failures or other severe problems within hours of a major upset.

Recommended analyses include these:

Primary On-Line

- Cation Conductivity
- Sodium*

Supplementary On-Line

- Dissolved Oxygen

Sodium monitoring is very effective for detecting condenser leaks. With a tight condenser, sodium levels in the condensate should be very low (<5 ppb), and in many cases less than 1 ppb. Cation conductivity can provide similar protection, although it is not quite as specific as sodium. Practical Example 6–1 illustrates an actual case history of a rather unusual condenser leak that plant personnel detected through on-line sodium analyses.

Dissolved oxygen analyses are very useful for monitoring air in-leakage to the condenser. Ideally, if the condenser air removal system is operating at maximum efficiency, dissolved oxygen levels should be below 7 ppb. Practically, 20 ppb is a good limit. A sudden increase in dissolved oxygen may indicate a problem at or near the condenser, which allows excess air to enter the system. I have illustrated several actual failure mechanisms in the section on condenser performance monitoring at the end of this chapter.

Practical Example 6-1

This incident occurred in an Admiralty-tubed condenser on an 80 MW coal-fired unit. Then and now, sodium concentrations in the condensate pump discharge typically average between 0.5 and 0.9 ppb. Once, during a three-week stretch of base-load operation, the sodium levels periodically increased to a range of 1 to 2 ppb, where they remained for several hours or perhaps up to a day before returning to normal. The sodium increase could be traced to any operational factors such as load change or soot blowing, so plant chemists concluded that a condenser tube leak had developed. Maintenance personnel did indeed find a pinhole leak in one tube. When they plugged the tube, the problem was cured. However, the question still remained as to why the contamination appeared and then disappeared with regularity. What seems most likely is that the hole frequently became plugged with debris that entered with the cooling water. On-line sodium analysis proved very sensitive in finding this small leak.

Condensate Polisher Effluent. Condensate polishers may be more or less complex depending upon the pressure of the boiler and whether steam is used to drive a turbine. Often, they are not used at low- and medium-pressure plants, or at plants that have no condensate return. We will examine monitoring for a complete system, however.

Recommended analyses for either a deep-bed or powdered-resin polisher include these:

Primary On-Line

- pH*
- Sodium*
- Silica*
- Cation conductivity

A condensate polisher is an on-line demineralizer, so instrument detection limits must be at low ppb levels or better. Sodium and silica analyses are very beneficial and will provide valuable data on performance. Cation conductivity is also useful for monitoring system conditions or bed exhaustion. Some deep-bed condensate polishers are operated in the ammonia form, in which the cation resin is first regenerated with acid and then is treated with an ammoniated solution to put ammonium ions on the exchange sites. Subsequent operation allows a gradual leakage of ammonia into the condensate, which conditions the pH. Ammonia can mask the leakage of other compounds, but this effect is cancelled out by cation conductivity monitoring.

Deaerator Inlet. In the electric utility industry, sampling of the deaerator inlet is important, especially if an oxygen scavenger is added at the condensate pump discharge to protect low-pressure heaters.

Recommended sample analyses include these:

Primary On-Line

- Dissolved oxygen*
- Oxygen scavenger*

Grab

- Iron
- Copper

Grab sample analyses for iron and copper provide data on condensate system corrosion and the performance of any treatment chemicals added to the condensate. For industrial condensate/feedwater systems without turbines, deaerator inlet sampling may be less critical due to the reduced possibilities of air in-leakage. In this case, periodic (daily or perhaps weekly) grab sample analyses for dissolved oxygen, when compared to grab sample deaerator outlet samples, are useful for monitoring the performance of the deaerator. Some of the more modern combined-cycle generating units may not even have a deaerator or the deaerator may be integral to the low-pressure boiler.

Deaerator Outlet. Deaerator outlet sampling for dissolved oxygen (grab is most common) helps the plant chemist monitor deaerator performance. An increase in D.O. levels over normal concentrations could indicate problems with the internal deaerator components. Trays may become misaligned. Or, a deaerator vent may have fouled with rust particles.

Feedwater or Economizer Inlet. This sample is very important, because this is the last checkpoint before the boiler itself. Feedwater chemistry can have a significant impact on boiler operation for several reasons. First, excessive feedwater contamination will reduce the boiler cycles of concentration and require increased blowdown. Second, improper control of feedwater chemistry may cause corrosion of feedwater piping and heat exchanger tubes, which will introduce iron oxide particles and copper corrosion products to the boiler. Third, in many steam generating systems feedwater is sprayed into the main steam (and reheat steam where applicable) for temperature control. Contaminants are directly introduced to the superheater and turbine via this process.

Recommended feedwater analyses include these:

Primary On-Line

- pH*
- Dissolved Oxygen*
- Oxygen Scavenger*
- Specific Conductivity*
- Cation Conductivity

Grab

- Ammonia
- Iron[1]
- Copper[1,2]

[1] *Weekly analysis as a minimum.*

[2] *Required only for units with copper-alloy tubed condensers and/or feedwater heaters.*

The optimum feedwater pH for systems containing mixed copper and iron metallurgy is 8.8 to 9.1, while for strictly iron-based systems the range is higher at 9.2 to 9.6. Dissolved oxygen levels below 5 ppb are recommended, however, a slight oxygen residual (2 ppb) should be maintained to preserve the magnetite layer (Fe_3O_4) that naturally forms on the metal surface. Several catastrophic pipe failures have been traced to excessive deoxygenation of the feedwater, which allowed erosion-corrosion of the base metal. This was outlined in chapters 4 and 5 in the sections on flow-assisted-corrosion. On-line monitoring of the oxygen scavenger hydrazine is possible, however, continuous monitoring for alternative scavengers may not be.

Ammonia sampling on a daily basis is important, especially if the feedwater system contains any copper-alloy materials. Ammonia in the presence of oxygen is a strong copper corrodent. A general rule of thumb states that ammonia concentrations should be maintained below 0.5 ppm.

Iron and copper analyses are quite useful in monitoring condensate/feedwater system corrosion. In a properly treated system, iron levels should be below 10 ppb (preferably 5 ppb) and copper concentrations less than 2 ppb.

Boiler Water. The boiler water sample is perhaps the most important of all, especially for drum-type boilers. This is due to high temperatures and the concentrating effect caused by recirculation of the boiler water. These factors greatly influence potential corrosion and scaling mechanisms. Furthermore, high concentrations of dissolved solids in the boiler water can introduce excessive contaminants to the steam, where they may form deposits and/or corrode superheater tubes and turbine blades.

Recommended boiler water analyses include these:

Primary On-Line

- pH*
- Specific Conductivity*
- Silica

Secondary On-Line

- Phosphate[1,2] *
- Sodium[2]

Grab

- Chloride
- Sulfate
- Ammonia[2]
- Dissolved Oxygen

[1] *For phosphate-treated units.*

[2] *For sodium-to-phosphate ratio calculations.*

The most important analyses are pH and silica. Boiler water pH must be maintained within a fairly narrow range (between 9 and about 10.5 depending on boiler pressure and design) to prevent corrosion of boiler tubes. Silica must be held below pressure-dependent limits, as silica will vaporize, carry over with steam, and precipitate in the turbine. This effect becomes very dramatic as pressure increases. For example, in a 900 psig boiler without reheat, the recommended maximum drum water silica concentration is 7 ppm. In a 2,400 psig boiler with reheat, the recommended maximum is 0.17 ppm!

In chapter 5 we looked at coordinated and congruent phosphate treatments. These have been very popular programs in drum-type boilers including HRSGs. An important requirement for coordinated or congruent phosphate treatment is sodium/phosphate ratio monitoring, because the ratio of sodium to phosphate in solution has a direct bearing on potential corrosion of the boiler tubes.

As we learned, if phosphate is the only major species in boiler water, the sodium-to-phosphate ratio can be calculated directly from phosphate concentration and pH. However, this direct calculation is inaccurate in systems where ammonia or amines are used for feedwater pH control. Yes, computer programs exist that take ammonia or amine concentrations into account and calculate a more accurate sodium-to-phosphate ratio. Personally, if I were designing a new monitoring system for a phosphate-treated unit, I would include both an on-line phosphate and sodium monitor to directly calculate Na/PO_4 ratios.

Chloride, sulfate, and sodium analyses are recommended at least on a grab sample basis. Sodium salts carry over with steam, and, if concentrated too heavily, cause corrosion of turbine blades. Ammonia, especially as concentrations increase, may exacerbate the carryover of chloride and sulfate.

Saturated Steam. Saturated steam analyses are useful for verifying that steam being produced by the boiler meets the turbine manufacturer's guidelines. As it turns out, main and reheat steam sampling are more accurate and more valuable, since such sampling can also detect impurities introduced to the steam from spray attemperators. Sometimes practicality plays a part in sample selection. When my former utility installed a comprehensive on-line system, the saturated steam sampling taps were already in place, but none existed for main or reheat steam. Installation of taps in a steam line requires a lot of work to maintain the integrity of the steam piping. Therefore, utility managers elected to only sample saturated steam and perhaps install main steam taps at a later date.

Recommended analyses include:

Primary On-Line

- Sodium*
- Silica*

These analyses provide a direct indication of steam purity. Both should be of very low concentration. In the past, turbine manufacturers have recommended a maximum silica limit of 20 ppb. Modern guidelines suggest 10 ppb for silica and 5 ppb for sodium in phosphate treated units.

Main Steam. Main steam samples are very important, particularly on units without reheat. Main steam analyses provide data on the purity of the product entering the turbine. These samples also may reveal contamination introduced through the attemperator.

Recommended analyses include these:

Primary On-Line

- Sodium*
- Silica*
- Specific and Degassed Cation Conductivity[1]

Supplementary On-Line or Grab

- Chloride[1]
- Sulfate[1]
- TOC[1]

[1] *For units without reheat.*

Sodium and silica analyses provide an excellent indication of steam conditions. Degassed cation conductivity is becoming a base-line measurement of steam purity. Chloride, sulfate, and TOC analyses allow the chemist to monitor carryover of potentially harmful corrodents to the turbine. Past recommendations by EPRI on steam turbine purity for units without reheat are shown in Table 6–1.

Reheat Steam. The hot reheat sample is equally or perhaps more important than the main steam sample in units with reheaters. This steam has already given up some of its energy in the high-pressure end of the turbine and is now entering the intermediate stage followed by the low-

EPRI Steam Turbine Purity Recommendations

Substance	Phosphate Treated Units Without Reheat (ppb)	All-Volatile units Without Reheat (ppb)
Sodium	10	5
Cation Cond.	0.35	0.25
Silica	20	10
Chloride	5	3
Sulfate	5	3
TOC	100	100

Table 6–1 *Steam Turbine Guidelines — 1*
Source: EPRI

pressure stage. It is in these stages when many impurities will drop out of the steam and collect on turbine blades.

Recommended analyses include these:

Primary On-Line

- Sodium*
- Silica*
- Specific and Degassed Cation Conductivity[1]

Supplementary On-Line or Grab

- Chloride[1]
- Sulfate[1]
- TOC[1]

[1] *Table 6–2 shows recommended steam purity limits for units with reheaters.*

Steam Purity in Heat Recovery Steam Generators. As we have already discussed, water quality in HRSGs should be consistent with guidelines established for conventional boilers. This helps maintain reliable

EPRI Steam Turbine Purity Recommendations
(with Reheaters)

Substance	Phosphate Treated Units With Reheat (ppb)	All-Volatile units With Reheat (ppb)	Once-through units All-Volatile units (ppb)
Sodium	5	3	3
Cation Cond.	0.3	0.15	0.15
Silica	10	10	10
Chloride	3	3	3
Sulfate	3	3	3
TOC	100	100	100

Table 6-2 *Steam Turbine Guidelines — 2*
Source: EPRI

performance of the unit. But what about steam? Where steam is supplying just the steam turbine in a combined-cycle unit, then the steam purity should be in line with the turbine manufacturer's guidelines. However, new developments in combined-cycle design may make such limits unusable. Researchers have found that steam is the most efficient coolant for components in combustion turbines that overheat unless cooled. This requires direct injection of the steam into the combustion turbine. Very pure steam is a requirement to prevent corrosion of the turbine blades. At the 1999 International Water Conference, a Siemens Westinghouse representative presented a paper on steam purity requirements for conventional turbines and combustion turbines. These are outlined in Table 6–3. As is clearly evident, the steam purity requirements for combustion turbines are much more stringent. Copper values are so low that this number may not be a realistic goal, at least in units that have copper-alloy condensers or heat exchanger tubes. This is obviously a factor that should be taken into design of the unit. If copper cannot be tolerated, then do not use copper-alloy condenser or feedwater heater tubes.

Steam Purity Requirements for Conventional and Combustion Turbines

Substance	Steam Turbine Limits (ppb)	Combustion Turbine Limits (ppb)
Sodium	5	1
Silica	10	30
Chloride	5	1
Sulfate	5	1
Sulfate	5	1
Phosphate	5	5
Iron	5	5
Copper	2	0.01

Table 6-3 *Steam Purity Requirements for Conventional Turbines and Combustion Turbines*
Source: International Water Conference

Developing Analytical Techniques

Two analytical methods that are gaining in popularity are dissolved hydrogen and oxidation-reduction potential (ORP) monitoring. As the reader may recall, the carbon steel in feedwater piping and waterwall tubes develop a protective coating of magnetite via the following reaction:

$$3Fe + 4H_2O \rightarrow Fe_3O_4 + 4H_2\uparrow \qquad [Eq.\ 6\text{--}1]$$

This reaction produces a background hydrogen concentration of a ppb or two in the steam. Several of the most common corrosion mechanisms also produce hydrogen. If plant personnel develop solid background data, then corrosion can be detected when hydrogen levels increase. Hydrogen monitors have been developed that detect ultra-low concentrations of the chemical. This is a promising technique for fine-tuning steam generation chemistry.

The reader may remember in chapters 1 and 2 the discussions about oxidation and reduction, and that some elements prefer to give up electrons in a chemical reaction and some prefer to accept electrons. Instruments are now available that measure this tendency of solutions. The technique has become popular for cooling water treatment, because solutions containing oxidizing biocides are easy to measure. Scientists are now trying to apply the technique to feedwater systems, as a monitoring and control device. As we learned in chapter 4, flow-assisted-corrosion is now recognized as a very serious issue in existing plants and HRSGs. The attack is in large part caused by reducing conditions that develop due to the use of oxygen scavengers. ORP monitoring offers promise as a monitoring and control technique for oxygen scavenger feed.

Instrument Maintenance

On-line instruments may be advertised as being rugged, put personally I have found them to be somewhat temperamental beasts that require regular calibration and periodic maintenance. Calibration frequency varies from instrument to instrument. Exposure of instruments to harsh conditions is not recommended. I have been to more than one plant where the original on-line system was set out on the coal-feeder floor or near the boilers. Naturally, very few of the instruments were still in operation. Where possible, it is best on install instruments in an enclosed room that has temperature control. Even a block building with a window air-conditioning unit helps a lot.

This comprises the basic list of recommended samples for a utility or large industrial steam generating system. Even more important than sample selection is how samples are extracted from the process streams. Unless samples are collected and conditioned properly, the data may be unrepresentative of actual conditions. Even the best instruments are worthless if the samples are not representative. The following section discusses this very important aspect of sampling.

Sample Extraction and Conditioning

An extremely important criterion in a monitoring program is sample extraction. Samples must be taken from the bulk solution. The techniques differ somewhat for liquid and steam sampling. Figure 6–2 shows a recommended liquid sample nozzle. The nozzle extends into the fluid and has a 45° beveled inlet that faces the flowing water. The nozzle is of sufficient length to prevent any interference from the laminar sub-layer that exists along the pipe surface.

Sampling of steam is a bit more complicated, for it is more difficult to extract a representative or homogeneous sample. One method to sample

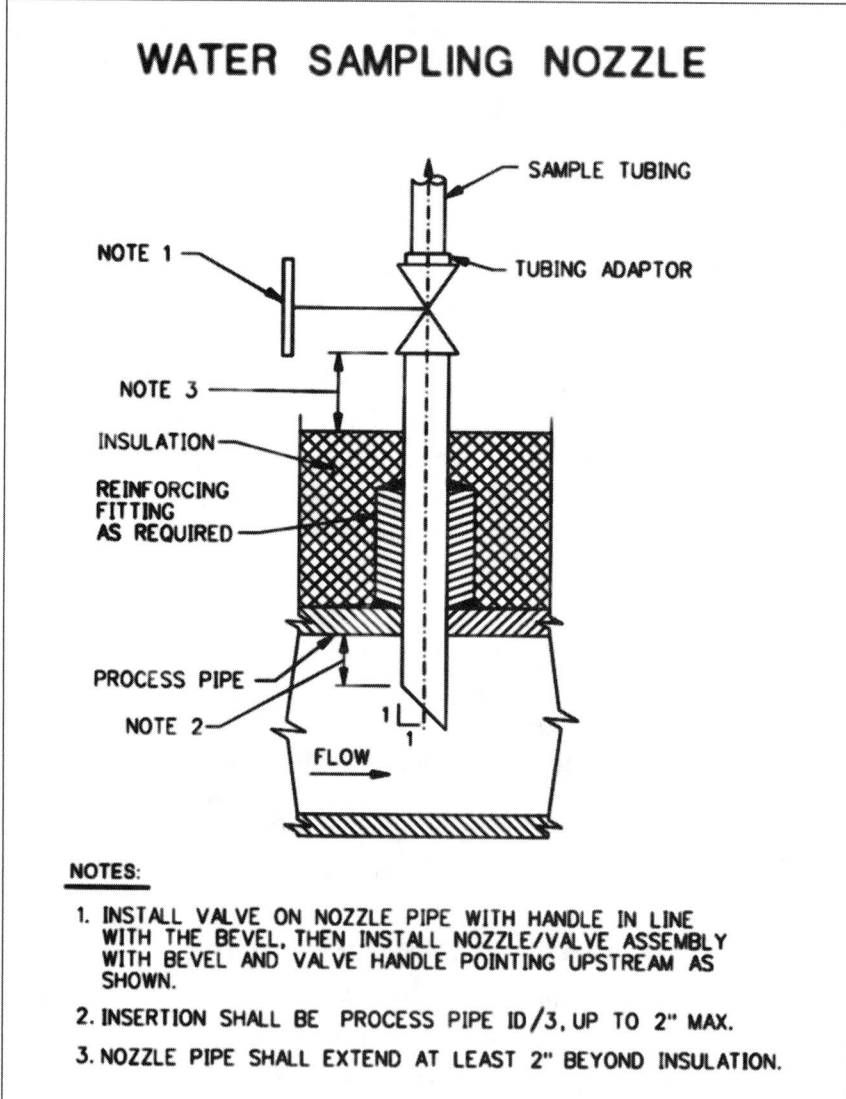

Fig. 6–2 *Recommended Liquid Sample Nozzle*

CHAPTER 6: MONITORING TECHNIQUES AND
CONTROL GUIDELINES

saturated steam is to actually install sample taps in the steam section of the boiler drum. The multiport sample tap (generically outlined in Figure 6–3, and installed in the steam outlet) is a common alternative. The multiport feature is designed to provide cross-sectional representation.

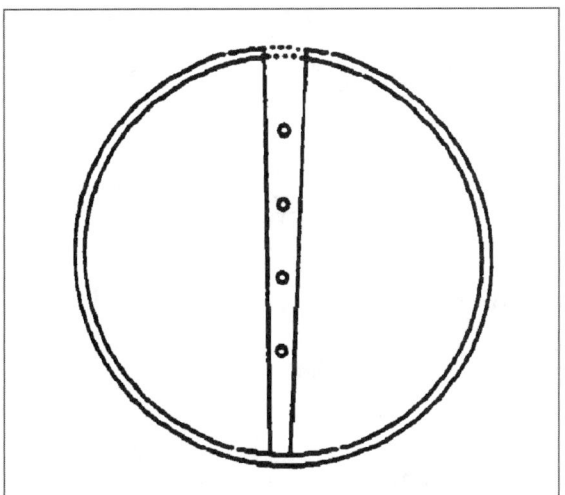

Fig. 6–3   *Multiport Steam Sample Tip*

Researchers have found that the best way to obtain a representative sample of main or reheat steam is to extract the steam at the same velocity at which it flows through the steam line. This is known as isokinetic sampling. Some debate still exists on whether samples can be completely extracted isokinetically, but the use of specially designed nozzles for this purpose is increasing. Figure 6–4 illustrates one of these nozzles. Nozzle design is by the Electric Power Research Institute and the nozzle is manufactured by Jonas & Consultants in Wilmington, Delaware.

Steam nozzles must be installed according to ASME boiler codes. These require precise welding techniques and heat treatment of the steam line. Retrofit of a steam-sampling nozzle into an existing line is not a simple procedure, but, given the improved accuracy that the nozzle provides, may easily pay for itself if even one steam system upset is prevented. I have not included procedures for welding, tapping, and heat treatment in this book, but any reliable sample nozzle or sampling system manufacturer will know where to obtain this information.

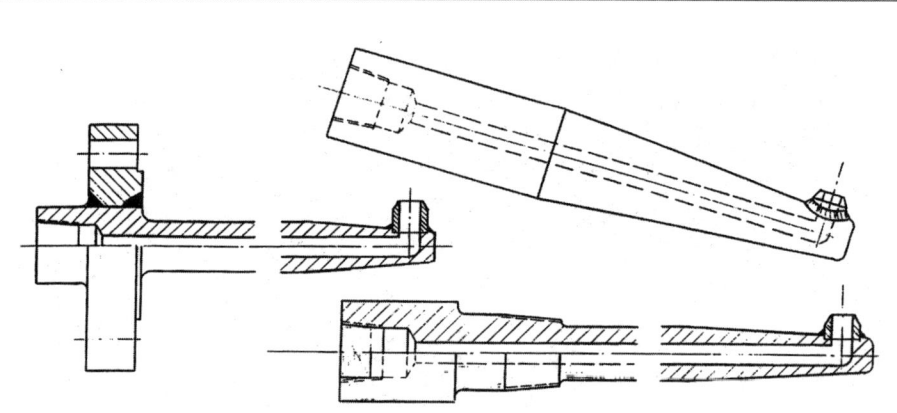

Three examples of EPRI Isokinetic Sampling Nozzles

Each nozzle is designed with considerations of vortex shedding, resonance, vibration, erosion, and strength of the attachment to the pipe.

MATERIALS:

Carbon Steel (C1018 or A105, as specified)
Stainless Steel (304 or 316, as specified)
Low Alloy Chrome Moly Steel (F11 or F22, as specified)
Other materials available upon request

PRESSURE - TEMPERATURE RATING
LBS. PER SQ. INCH

MATERIAL	TEMPERATURE - °F.						
	70°	200°	400°	600°	800°	1000°	1200°
Carbon Steel	5950	5750	5450	5250	4000	1750	—
A.I.S.I. 304	7800	7050	6400	6150	6000	5190	1875
A.I.S.A. 316	7800	7800	7250	7100	6950	5800	2720
F-11	7350	7350	7350	7350	7350	2898	504
F-22	7224	7224	7224	7224	7098	3192	546

Ordering Information:

1. Pressure, temperature, and mass flow rate of the sampled fluid.
2. Pipe ID, wall thickness, and material.
3. Desired sample flow.
4. Attachment to the pipe: weld, thread plus seal weld, flange, etc.
5. Nozzle material.
6. Thickness of thermal insulation.

Fig. 6–4 *Nozzle for Isokinetic Steam Sampling*

Sample point location is also very important. Ideally, boiler water sample taps should be installed in a downcomer line, although the continuous boiler blowdown is a common alternative. The ASTM recommends that other liquid sample taps be installed at least 25 pipe diameters downstream of a chemical injection point if the flow is turbulent and 50 pipe diameters if the flow is laminar. A rule-of-thumb guideline suggests that the tap be located at least 10 pipe diameters downstream of a fitting or flow disturbance. Where possible, taps should be located in long vertical runs. When these are unavailable, a long horizontal run is the next best choice. Jonas & Consultants recommend that steam taps be installed 35 pipe diameters downstream and 4 pipe diameters upstream of any flow disturbances, and at a 12 o'clock position. For liquid samples in horizontal runs, the literature has recommended both the 3 o'clock and 1:30 positions for sample tap location.

After the sample has been extracted, it must be conditioned (temperature and pressure reduction) for analysis by the on-line instruments. The proper flow rate must also be established so that dissolved or suspended solids do not precipitate on the sample tube walls, or conversely that the sample does not pick up deposits that have already attached to the sample tube walls. For years, the literature has recommended that steam samples be cooled and condensed shortly after sample extraction to preserve sample integrity. As steam passes through a sample line the temperature decreases and the density increases. Too much natural cooling (without condensation to a liquid) was thought to affect the solubility of dissolved contaminants and cause some deposition onto the tube walls. Thus, an unconditioned steam sample that passes through a sample line from the source to the analyzer may not represent actual process conditions.

The method developed to prevent this requires conversion of the steam to condensate immediately after the sample is extracted. Figure 6–5 outlines this arrangement. The sample cooler is placed directly after the root valves and condenses the steam before any dissolved solids begin to precipitate. Originally, this design included condensate recirculation, in which a portion of the condensed steam is returned to the cooler inlet. Some experts thought that this improved sample integrity, and this design can be found in ASTM guidelines. However, practical operating experi-

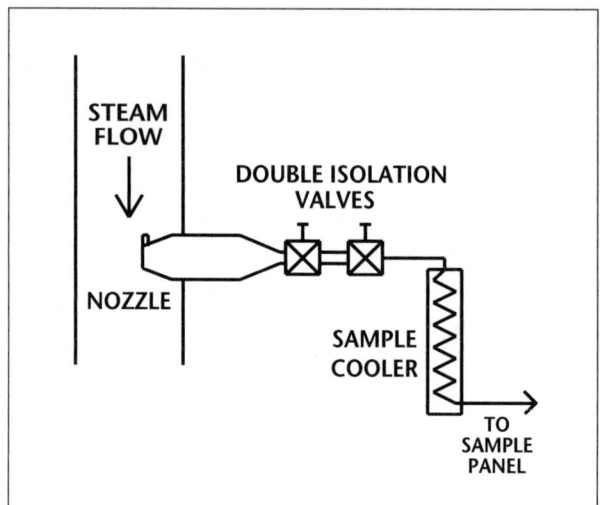

Fig. 6–5 *Converting Steam to Condensate*

ence has shown this to be very unmanageable, and the simpler system shown in Figure 6–5 is more common.

Recent tests conducted by personnel at a major electric utility indicate that remote cooling of steam samples may not be as critical as has been thought. The plant staff found that side-by-side tests of remotely cooled samples versus those brought hot to the sample rack showed very little difference in chemistry. The sample lines were uninsulated and approximately 150 feet in length.

Conversely, in a recent report from another plant, the chemists found that sintered wire filters placed in sample lines to protect instruments were actually affecting sample integrity, which led to false readings. Both of these last two examples indicate the some of the variables that one must consider when monitoring water/steam chemistry.

Flow rate of the sample is also important in preserving sample integrity. High or low flow rates can cause changes in the sample constituency. The recommended guidelines for sample linear flow rates are as follows:

- Liquid — 5 to 6 feet per second
- Steam — 36 feet per second

CHAPTER 6: MONITORING TECHNIQUES AND
CONTROL GUIDELINES

Sample line sizes of $1/4$" o.d. and $3/8$" o.d. are most common. Volumetric flow rates in these lines to maintain a 6 foot-per-second (fps) linear flow rate are 1200 cc/min and 3300 cc/min, respectively. In some cases, the 6 fps linear flow rate taxes the sample conditioning equipment. Sampling experts have found that flow rates can possibly be lowered to 3 fps without seriously compromising sample integrity. The guideline of 36 feet-per-second for steam was established at a baseline pressure of 1500 psig. Higher flow rates are recommended for lower pressures and lower flow rates at higher pressures. Where steam samples are cooled directly after extraction, liquid velocities apply.

In summary, the following list will serve as a general guideline for designing and installing sample lines.

- Select a suitable material for the sample lines. Pick something that will not corrode easily and introduce particles to the sample. 316 stainless steel is usually a good choice. Carbon steel and copper are bad.

- Size the line diameter to match recommended flow rates. Lines that are too small or too large will harm sample integrity.

- Keep sample lines as short as possible.

- Minimize the number of elbows or flow disturbances in the sample line.

- Cool steam samples at the source. This may require fairly long plumbing runs for the cooling water, and perhaps a booster pump, but these are not difficult tasks.

- Make sure that horizontal runs of the sample line slope downwards towards the sample conditioning rack. This prevents moisture and deposits from collecting in the lines.

Final Sample Conditioning. Primary sample coolers should be designed to reduce the sample temperatures to around 100°F. At present, some debate exists about the need for additional cooling. Many analytical instruments can be equipped with temperature compensating devices that will allow them to analyze samples at temperatures at or a bit above

100°F. However, the accuracy of the analyses at these temperatures is questionable. More precise readings are obtained if the sample is cooled to 77°F (25°C), ±0.5°F. This is the standard temperature for analytical readings. Accordingly, where possible a water/steam sampling system should be equipped with a secondary cooling system to bring sample temperatures down to 77°F. Even if this temperature drifts slightly, temperature-compensating devices on any instrument should work better than if the sample was still at 100° or so.

Economically it is often most sensible to route multiple samples to a central location for conditioning and analyses by the on-line instruments. Conditioning takes place in a device appropriately known as the sample conditioning rack. Figures 6–6 and 6–7 show front and back views of a conditioning rack. Figure 6–8 outlines a common conditioning configu-

Fig. 6-6 *Conditioning Rack (Front)*
Courtesy of Sentry Equipment Corp., Oconomowoc, WI

ration for one sample in the rack. Each component provides an important function. The high-pressure reducer, typically a rod-in-tube device, conditions even the highest-pressure samples and reduces the pressure to near instrument requirements. Following the pressure regulator is a high-temperature shutoff solenoid (HTSS). This device will close if it detects sample temperatures above pre-set values. Ideally, the HTSS would be

CHAPTER 6: MONITORING TECHNIQUES AND
CONTROL GUIDELINES

Fig. 6–7   *Conditioning Rack (Rear)*
Courtesy of Sentry Equipment Corp., Oconomowoc, WI

located directly downstream of remote primary coolers, so that loss of cooling water would cause the solenoid to stop sample flow before the primary cooler became too hot. However, this location is often not possible or practical. Sample coolers are designed with pressure relief valves to protect the cooler if cooling water flow is interrupted.

Following the HTSS is a low-pressure blowdown line. This allows the operator to bypass the sample around most of the conditioning equipment during unit startups. This is necessary because startups introduce suspended solids to the samples, which can clog downstream instruments. Next in the conditioning network is the secondary cooling system, which uses chilled water to cool the samples to 77°F. Two methods of secondary cooling are possible. Either chilled water can be routed to individual sample coolers, or the samples themselves can all be routed to an isothermal bath. Some experts believe that the former arrangement allows for more precise temperature control.

Following the secondary cooler are traditional flow control devices including a total flow rotameter, pressure gauge, temperature gauge, and

Fundamentals of Steam Generation Chemistry

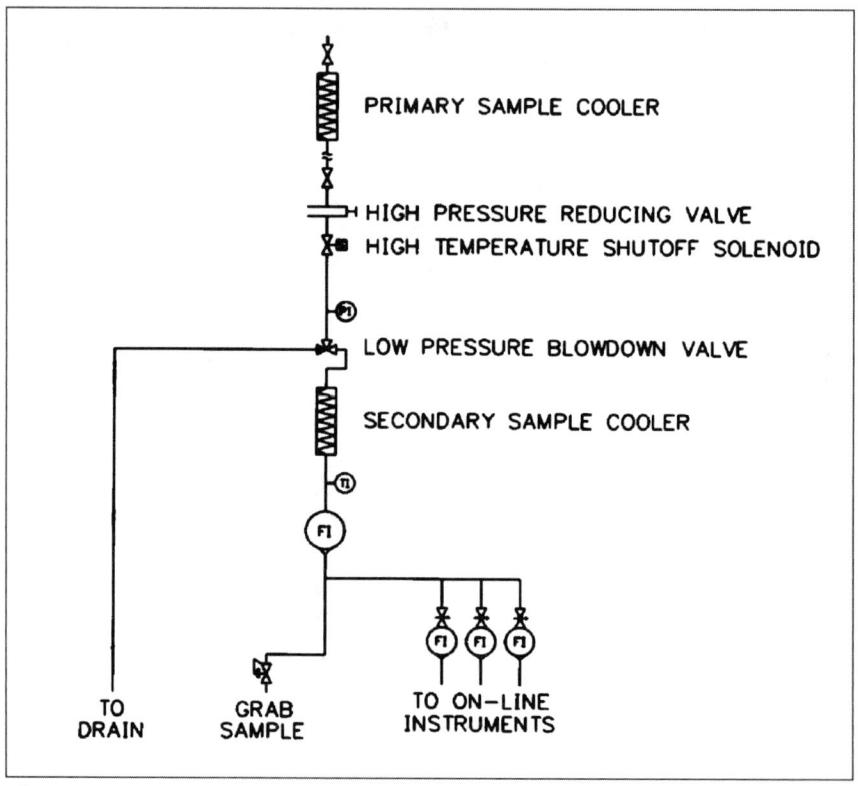

Fig. 6–8 *Conditioning Configuration*

individual rotometers for the on-line instruments. These give the chemist or operator additional flexibility in controlling feed to the instruments and grab sample port.

Figure 6–9 shows the latest design of a sample-conditioning network, with a controller added. The controller monitors total flow rate and automatically adjusts the pressure-reducing valve to maintain constant flow. This feature merits strong consideration. Changes in flow will upset the equilibrium of a sample, which can cause analytical errors. It may take many hours to perhaps a day for equilibrium to return, during which time the flow may have changed again.

Instruments and Communications. On-line instruments are available from a wide variety of manufacturers. Virtually all of the analyses listed

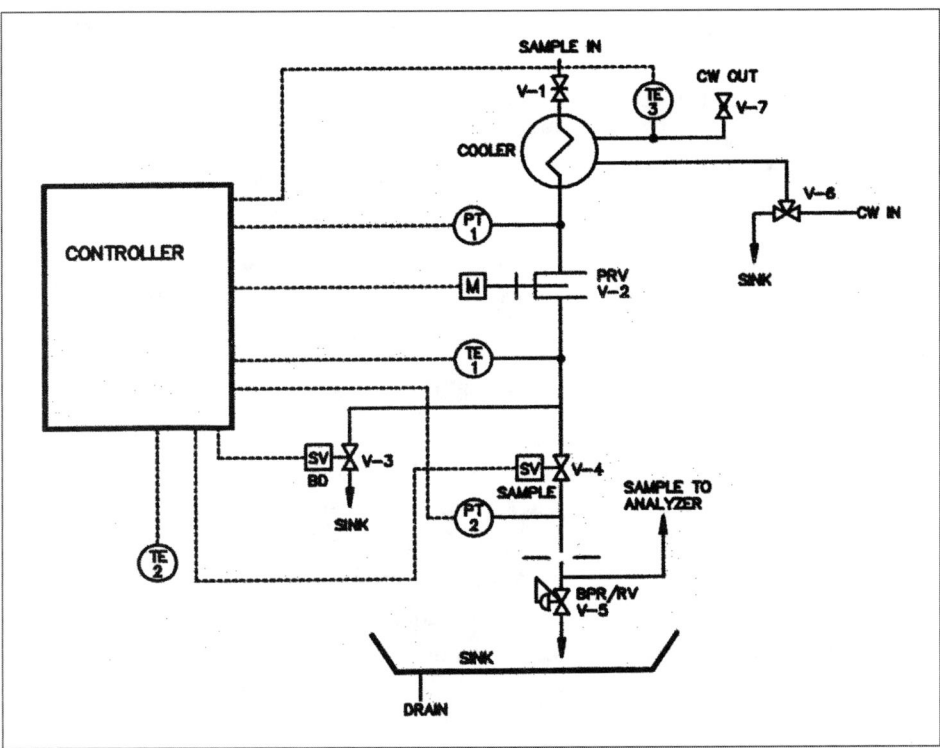

Fig. 6-9 *Conditioning System with Controller*
Courtesy of Sentry Equipment Corp., Oconomowoc, WI

in the previous sections can be performed on-line. A critical feature of on-line monitoring is communication of data to plant personnel so that any upsets can be detected immediately. While on-line monitors are or can be equipped with local alarms, virtually all will provide an electronic output that can be directed to a personal computer, programmable logic controller (PLC), or distributed control system (DCS). Thus, it is relatively simple to send data to such areas as the plant control room and engineering offices. Practical Example 6–2 outlines an on-line monitoring system that has provided useful results.

Practical Example 6-2

Remember the chemistry upset outlined in Practical Example 4-3? I was given responsibility for the installation and programming of the on-line monitoring system to prevent such surprises from happening again. Due to a limited budget, our consultant and lab manager selected the samples and instruments listed below for each of the three main units:

- Condensate Pump Discharge — Sodium, cation conductivity.
- Deaerator Inlet — Dissolved oxygen, hydrazine.
- Feedwater/Economizer Inlet — Dissolved oxygen, pH, hydrazine, specific conductivity.
- Boiler Water — Specific conductivity, pH, phosphate, silica.
- Saturated Steam — Sodium, silica.

The system has performed well, especially in detecting condenser leaks. The system detected the leak in Practical Example 6–1. All data from the instruments is sent to a PLC, from which data is collected by remote PCs located in the plant control room, engineering offices, and main laboratory. Thus, all critical personnel have access to real-time data. The original software included a trending program that allows up to 24-hour's worth of data to be shown on the remote screens. The software also came equipped with a datalogging program that downloads readings at user-selected intervals into common spreadsheet programs.

Before ending this chapter, I wish to discuss condenser performance monitoring. Upsets in condenser performance can be the first indicators of cooling system fouling or air in-leakage. While both of these conditions impair heat transfer in the condenser, they may also be the precursors to corrosive conditions either on the cooling water or steam side of the unit.

Condenser Performance

Many of the conditions that cause corrosion or other difficulties in a condenser also affect condenser performance. Scaling or fouling of condenser tubes restricts heat transfer. Excess air in-leakage does the same because air coats the tubes. Waterside fouling or air blanketing may so affect heat transfer in the condenser that the unit loses 1% or more of overall heat rate. For a large unit, this can add up to some serious money over a year.

Various techniques are available to evaluate condenser performance. For instance, the discharge line from the condenser air-removal section can be equipped with a bypass loop containing a flow meter. Plant operators can and should routinely check the air flow from the condenser. The literature suggests that the flow rate should average about 1 standard cubic foot per minute (SCFM) per 100 MW of output. However, conditions vary from unit to unit. More important is to establish baseline data during normal operation and then watch for changes.

A common technique to evaluate condenser performance is monitoring of the cooling water inlet, cooling water outlet, and condensed steam temperatures. Changes in these values, especially those of the Terminal Temperature Difference (TTD, the difference between the condensed steam temperature and cooling water outlet) may reflect changes in condenser performance. Consider the data shown in Table 6–4, which is taken from the log sheet of an actual condenser. The tubes were fairly clean when the first set of data was collected, but were fouled with microbiological deposits at the time of the second reading. The cooling water inlet temperature during the second reading is higher due to sum-

Condenser Data
Reflecting Performance Loss Due to Fouling

Parameter	July 11th, 1995 Tubes Clean	August 16th, 1995 Tubes Fouled
Unit Load (MW)	185	190
Cooling Water Inlet Temperature (°F)	80.0	88.5
Cooling Water Outlet Temperature (°F)	96.0	106.5
Condensed Steam Temperature (°F)	106.0	128.0
TTD (°F)	10.0	21.5

Table 6–4 *Condenser Performance Data*

mer warming. This in itself would have an effect on outlet and condensed steam temperatures, but the important item to note is the change in the TTD. It is obviously much greater. This was due to the reduction in heat transfer caused by the deposits.

Another technique is monitoring of condenser backpressure. Tube fouling or air in-leakage will cause an increase in backpressure due to reduced heat transfer. However, an increase in cooling water inlet temperature also increases backpressure, even though the condenser may be operating perfectly. Condenser performance problems may go undetected because changes in fluid temperatures are often gradual. Several computer programs exist to accurately track condenser performance. One such program was developed at my former utility. I first reported on it in 1992 in Power Engineering magazine. The original BASIC version of the program is outlined in Learning Aide 6–1, and is based on information supplied by the General Physics Corporation in one of their training classes. They graciously allowed me to report on it in the aforementioned Power Engineering article.

The program requires 13 simple-to-obtain inputs.

- Cooling water inlet temperature.
- Cooling water density.
- Cooling water outlet temperature
- Condensed steam temperature
- Cooling water flow rate
- Circulating water correction factor
- Condenser tube correction factor
- Number of condenser tubes
- Number of tube passes
- Inside tube diameter
- Outside tube diameter
- Tube length
- A constant, C

CHAPTER 6: MONITORING TECHNIQUES AND
CONTROL GUIDELINES

Because the tube dimensions for any condenser are constant, the program can be simplified by including this data in the logic. Thus, the number of inputs can be significantly reduced if constants for tube length, i.d., o.d., number of tubes, and number of passes are written into the program. In addition, the density of water decreases by only 0.5% over the temperature range of 40°F to 90°F, so a constant value for water density can be used without noticeably affecting the calculation. Guidelines for selecting C are listed in the program, and those for condenser tube correction factor and cooling water correction factor in the attachments to the Learning Aide (Table 6–5 and 6–6). These values help the program account for changes in condenser design and changes in cooling water inlet temperature. For any particular condenser, C and the condenser tube correction factor are constant, and may be permanently placed in the program. This program can be easily written into a spreadsheet form, with a log sheet directly included.

The program calculates an ideal heat transfer coefficient (U_i), which it then compares to a calculated actual heat transfer coefficient (U_a). The inverse of this value multiplied by 100 is known as the cleanliness factor. When condenser tubes are placed in service, they quickly develop an oxide coating. This coating, which actually protects the metal substrate, retards heat transfer. Thus, a condenser free from mineral or microbiological deposits (or excess air in-leakage) will still only achieve about 85% of the ideal heat transfer. A cleanliness factor of 85% indicates good performance.

An important issue with this program is accurate temperature readings. The literature suggests that inlet and outlet cooling water may stratify and give different temperature readings at the pipe wall than in the center. This will affect absolute values, although the program may still successfully observe trends. More consistent results are possible when readings are taken at high loads.

I successfully used the program to monitor the performance of three utility condensers over a five-year period. The program will detect changes in condenser performance due to scaling, microbiological fouling, or excess air in-leakage. For example, the condenser operating data outlined in Table 6–4 gave a 70.3% cleanliness factor for the first set of data and a 43.7% for the second set. This was consistent with the change in TTD and before-and-after visual observations.

The following Practical Examples outline several actual histories of the program in use. Two of the examples show the effect of air in-leakage on condenser performance.

Practical Example 6-3

I had been performing thrice-weekly cleanliness factor analyses on the utility's largest condenser, which handled a steam flow of at 1,000,000 lb/hr. The values remained very steady in the mid-70% range for several months, but suddenly within two days dropped to 45%. Waterside fouling does not occur this rapidly. Such drastic changes are more indicative of excess air in-leakage. The maintenance department was so notified. When maintenance personnel inspected the condenser, they discovered a crack in the condenser shell where a heater drips line penetrates. Once they sealed this crack, the cleanliness factors returned to previous values where they remained for another two months until suddenly dropping again. The seal had failed. The maintenance crew then welded a collar around the drips line, which totally sealed the crack and cured the problem.

Practical Example 6-4

In another instance, I had been collecting thrice-weekly readings on two, 690,000 lb/hr condensers. Suddenly, one condenser began performing erratically. At maximum unit loads, the cleanliness factors ranged between 70% to 75%, but at low loads the factor dropped as low as 18%. Again, such fluctuations could not have been the result of waterside fouling. Utility managers brought in a leak detection firm to look for air leaks. The inspectors used helium leak detection to completely check the condenser and low-pressure end of the turbine. They classified leaks as large, medium, and small, and found over a dozen leaks, including two large ones, one of which was caused by a crack in the expansion joint between the turbine exhaust and condenser. Maintenance crews repaired all leaks, but this did not solve the problem. Finally, an operator discovered that a trap on a line from the gland steam exhauster was sticking open at low loads. The trap and line are designed to return condensed gland steam from the condensate subcooler to the condenser, but vent gases to the atmosphere. When the trap stuck open, the strong condenser vacuum pulled outside air in through the vent. Once maintenance personnel replaced the trap, the condenser performance problems disappeared.

Practical Example 6-5

The last example illustrates how the program detected a problem that had never occurred before. The 1,000,000 lb/hr condenser had been in operation for 10 years but had never suffered from scaling. During one very dry summer, the lake volume decreased dramatically, and lab chemists calculated that the dissolved solids concentration in the lake increased 400% over normal values. However, no thought was given to the possibility of scale formation. Throughout the summer the cleanliness factor declined slowly but noticeably from around 80% to 45%. When the unit came off line for an autumn outage, an inspection team found that the waterside of the tubes was completely covered with a layer of calcium carbonate, less than one millimeter in thickness. The deposits were a direct result of the drought. Plant personnel observed an interesting peculiarity at this time. The condenser that scaled is equipped with 90-10 and 70-30 copper-nickel tubes. The two other condensers, both tubed with Admiralty, did not show scale buildups, even though operating temperatures were similar. The different metallurgy may have been the reason.

The program generates similar data when condenser tubes accumulate microbiological deposits. The program can be very useful for detecting the onset of microbiological fouling, and for scheduling shock chlorine treatments. A word of warning though, once microbiological colonies become established, shock chlorination does not always completely clean the tubes. One year, when I was monitoring performance of the 1,000,000 lb/hr condenser, the cleanliness factor dropped from around 80% in the early spring to 40% by early summer. Past experience suggested microbiological fouling, which turned out to be the case. In midsummer, operating personnel shock chlorinated the condenser, but this only restored the cleanliness factor to around 65%. When a maintenance crew opened the condenser during an autumn outage, many slime deposits were still evident. The maintenance crew had to mechanically scrape the condenser to remove deposits. This example points out the tenacity of microbiological deposits.

Conclusion

This article outlines essential ideas, guidelines, and equipment needed to accurately analyze steam generating system samples. The importance of good sampling cannot be overemphasized. When compared to the possible loss of revenue due to a forced outage, representative sampling makes perfect sense.

Learning Aide 6-1

CONDENSER CLEANLINESS MONITORING PROGRAM

The folowing computer program, written in BASIC, determines the cleanliness of a condenser, expressed as a percentage. The calculation is made by determining the heat transfer coefficient of the condenser, and then comparing it with the ideal heat transfer coefficient for that condenser.

Following the program code are reference tables of condenser tube and cooling water correction factors.

```
10   CLS: LOCATE 5,1
20   PRINT "Condenser Cleanliness Factor Program"
30   PRINT
40   PRINT "Enter the circulating water inlet temperature (F) "
50   INPUT TIN
60   PRINT
70   PRINT "Enter the density of the inlet cooling water (lb/ft^2) "
80   PRINT "(a value of 62.3 is sufficient for all conditions) "
90   INPUT RHO
100  PRINT
110  PRINT "Enter the circulating water outlet temperature (F) "
120  INPUT TOUT
130  PRINT
140  PRINT "Enter the condenser steam temperature (F) "
150  INPUT TSAT
```

```
160  PRINT
170  PRINT "Enter the circulating water flow rate (GPM) "
180  INPUT FLOW
190  PRINT
200  PRINT "Enter the circulating water correction factor as"
210  PRINT "outlined "
220  INPUT CWCF
230  PRINT
240  PRINT "Enter the condenser tube correction factor as"
250  PRINT "outlined "
260  INPUT CTCF
270  PRINT
280  PRINT "Enter the number of condenser tubes "
290  INPUT NT
300  PRINT
310  PRINT "Enter the number of condenser tube passes "
320  INPUT NP
330  PRINT
340  PRINT "Enter the inside tube diameter (in) "
350  INPUT ID
360  PRINT
370  PRINT "Enter the outside tube diameter (in) "
380  INPUT OD
390  PRINT
400  PRINT "Enter the tube length (ft) "
410  INPUT L
420  PRINT
430  PRINT "In the following calculations, the ideal heat transfer"
440  PRINT "is determined through the use of an empirical constant"
450  PRINT "determined by the Heat Exchange Institute. This constant,"
460  PRINT "known as C, varies depending on the outside tube diamter."
470  PRINT "The values of C are as follows:"
480  PRINT
490  PRINT "  5/8 and 3/4 inches, C=267"
500  PRINT "  3/4 and 7/8 inches, C=263"
510  PRINT "1-1/8 and 1-1/4 inches, C=259"
```

Chapter 6: Monitoring Techniques and Control Guidelines

```
520  PRINT "1-3/8 and 1-1/2 inches, C=255"
530  PRINT "1-5/8 and 1-3/4 inches, C=251"
540  PRINT "1-7/8 and 2 inches, C=247"
550  PRINT
560  PRINT "Enter a value for C "
570  INPUT C
580  PRINT
590  REM  Q = Heat Transfer (BTU/hr)
600  REM  DTLM = Log Mean Temperature
610  REM  SAOD = Surface Area of Outside Tube Surface (ft^2/ft)
620  REM  UA = Actual Heat Trasnfer Coefficient
630  REM  VL = Linear Velocity of Water Through the Tubes (ft/sec)
640  REM  UI = Ideal Heat Transfer Coefficient
650  REM  UD = Design Heat Transfer Coefficient
660  REM  CF = Cleanliness Factor
670  Q=8.0203*RHO*FLOW*(TOUT-TIN)
680  DTLM=(TOUT-TIN)/LOG((TSAT-TIN)/(TSAT-TOUT))
690  SAOD=3.14159*(OD/12)
700  UA=Q/(DTLM*SAOD*L*NT)
710  VL=(.002228*FLOW)/((((ID/24)^2)*(3.14159*NT/NP))
720  UI=C*SQR(VL)
730  UD=UI*CWCF*CTCF
740  CF=(UA/UD)*100
750  PRINT "The Cleanliness Facter = ";:PRINT USING"###.#";CF
```

Condenser Tube Correction Factors

Tube Material	Tube wall Gauge — BWG						
	24	22	20	18	16	14	12
Admiralty Metal, Arsenical Copper, or Copper Iron 194	1.06	1.04	1.02	1.00	0.96	0.92	0.87
Aluminum Brass or Aluminum Bronze	1.03	1.02	1.00	0.97	0.94	0.90	0.84
90–10 Cu–Ni	0.99	0.97	0.94	0.90	0.85	0.80	0.74
70–30 Cu–Ni	0.93	0.90	0.87	0.82	0.77	0.71	0.64
Cold Rolled Low Carbon Steel	1.00	0.98	0.95	0.91	0.86	0.80	0.74
Stainless Steels Type 304/316	0.83	0.79	0.75	0.69	0.63	0.56	0.49
Titanium	0.85	0.81	0.77	0.71	—	—	—

Table 6–5 *Condenser Tube Correction Factors*
Source: Heat Exchange Institute

Cooling Water Correction Factors

Inlet Temperature (°F)	CWCF	Inlet Temperature (°F)	CWCF
30	0.550	65	0.960
31	0.562	66	0.970
32	0.574	67	0.978
33	0.586	68	0.986
34	0.601	69	0.993
35	0.615	70	1.000
36	0.628	71	1.005
37	0.641	72	1.010
38	0.655	73	1.015
39	0.668	74	1.020
40	0.683	75	1.025
41	0.694	76	1.029
42	0.707	77	1.033
43	0.720	78	1.037
44	0.733	79	1.041
45	0.747	80	1.045
46	0.760	81	1.048
47	0.772	82	1.051
48	0.785	83	1.054
49	0.797	84	1.057
50	0.810	85	1.060
51	0.822	86	1.063
52	0.833	87	1.066
53	0.844	88	1.069
54	0.855	89	1.072
55	0.865	90	1.075
56	0.875	91	1.078
57	0.885	92	1.080
58	0.895	93	1.083
59	0.905	94	1.085
60	0.915	95	1.088
61	0.925	96	1.090
62	0.934	97	1.092
63	0.942	98	1.095
64	0.951	99	1.097
		100	1.100

Table 6–6 *Cooling Water Correction Factors*
Source: Heat Exchange Institute

COOLING WATER CHEMISTRY 7

Cooling of turbine exhaust steam or process fluids is a critical function at electric utilities and industrial plants. The cooling water flow rates required for these processes, especially steam condensation, are quite high, which prohibits the purifying techniques (demineralization, RO) that are common for smaller operations such as production of boiler makeup. Yet, contaminants can cause severe problems in cooling water systems if not properly controlled. This chapter outlines cooling water chemistry and treatment methods. Changing environmental regulations and improvements in cooling water treatment chemicals have added a complexity to treatment selection.

Cooling Systems

Most cooling systems belong to one of the following categories:

- Once-through
- Open recirculating (evaporative cooling towers)
- Closed recirculating

Each has special requirements for water treatment. Let's look at some of the basics of these systems.

Once-Through Cooling

Once-through systems are common at electric utilities or other power generation facilities. Flow rates may range from tens of thousands to hundreds of thousands of gallons per minute, or more. The source of supply is a lake, river, or the ocean due to the large quantities of water needed.

The greatest advantage of once-through cooling is that the water does not become concentrated as it does in a cooling tower, and thus has a much lower scale-forming potential. Even so, some surface supplies, especially rivers, exhibit changing water qualities due to seasonal effects or other factors. The primary disadvantage of once-through cooling is that the effluent returns to the source at a significantly higher temperature. Increasingly stringent regulations on thermal discharge, and its effects on aquatic organisms, are restricting new construction of once-through systems, and in some cases they may not be allowed.

Chemical treatment for once-through systems is usually fairly simple. Often, only a good biocide program is necessary. We will cover this in more detail later.

Open Recirculating Systems

In an open recirculating cooling system, water travels through the condenser or heat exchangers, to and through a cooling tower, and back again. High-volume flow rates are also possible with open recirculating systems, but with much less water discharge to the environment. The heart of an open system is the cooling tower (Figure 7–1), which cools water by cascading it through air. Depending upon air temperature and humidity, 65% to 90% of the heat transfer occurs through evaporation of water. Modern cooling towers can basically be divided into two categories, mechanical draft (air forced through the tower) and natural draft (air flows naturally through the tower).

Mechanical draft towers can, in turn, be divided into two other categories, crossflow and counterflow (single or dual entry), in which the air is either pushed (forced-draft) or pulled (induced-draft) through the tower. Operation is relatively straightforward. Warm cooling water is sprayed into flowing air, which absorbs heat. The warmed air is ejected from the tower,

CHAPTER 7: COOLING WATER CHEMISTRY

Fig. 7-1   *Cooling Tower*
Photo Courtesy of Marley Cooling Tower, Overland Park, KS

while the cooled water falls to a basin where it is recirculated. The falling water is broken up by fill material in the tower to enhance air/liquid contact and heat transfer. Most of the heat transfer occurs through evaporation of a small fraction of the cooling water. The remainder is sensible heat transfer from the water to the moisture-laden air stream.

For very large cooling water systems, natural-draft, counterflow towers are common. These are typified by the several-hundred-foot tall hyperbolic towers that one sees at a large utility, and that many citizens often mistake for nuclear reactor buildings. Hyperbolic towers generate a chimney effect to induce air flow. As the air warms during its contact with the water, it rises and flows out of the tower, pulling in cooler air from below.

One of the most important components of any cooling tower is fill. Fill comes in many shapes and patterns. Wooden splash bars were once the primary material, but these have mostly been supplanted by more sophisticated designs and materials, as is illustrated in Figures 7-2 and 7-3. These newer materials include vertically-aligned sheets of corrugated plastic known as film

Fig. 7–2 *Fill Material in Cooling Towers — 1*
Photo Courtesy of Marley Cooling Tower, Overland Park, KS

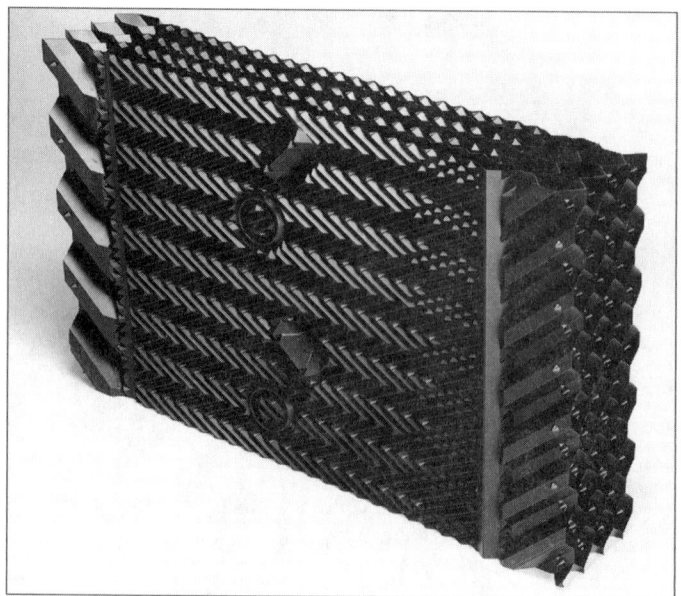

Fig. 7–3 *Fill Material in Cooling Towers — 2*
Photo Courtesy of Marley Cooling Tower, Overland Park, KS

fill, which is commonly used in counterflow towers. Film fill, while being very efficient, can be problematic if microbiological fouling or scaling is not carefully controlled, as the close spacing between layers makes the material susceptible to plugging. In fact, the increasing use of film fill has been one of the driving factors toward improvements in treatment programs.

The number of times that water is concentrated in the cooling tower is known as the cycles of concentration (C). C is usually determined by comparing the concentration of some very soluble ion, such as chloride or magnesium, in the makeup to that in the recirculating water. For example, consider a tower with a makeup chloride level of 50 ppm. If the chloride concentration in the recirculating water is 300 ppm, then C is 6.0. Specific conductivity may also be used for this determination, although conductivity is not completely linear with increasing dissolved solids. Conductivity can also be affected by treatment chemicals added to the water.

The maximum allowable cycles of concentration is site specific, and depends on the effectiveness of corrosion and scale inhibitor programs and also on the quality of the makeup to the tower. Naturally soft water or softened supplies may allow for increased cycles of concentration due to the reduced scaling potential, although softening may increase the corrosion potential. Cycles must be kept low when the makeup supply is hard or highly ionic. Seawater is a prime example of the latter. Cycles of concentration in a seawater-cooled tower may be limited to less than 2 due to the high ionic concentration of the recirculating water and its potential for corrosion. Conversely, cycles of 20 or even higher are possible with treated or soft makeup water. Generally, six to eight cycles of concentration is a reasonable range.

Regardless of the quality of the makeup water or effectiveness of a corrosion or scale inhibitor program, cycles of concentration have a limit in any system. Some water must be continually removed to prevent excessive buildups of dissolved solids. This is known as blowdown. An additional, very small blowdown occurs due to entrainment of water droplets in the air exiting the tower. This is known as drift. Drift typically ranges from about 0.3 to 0.05 percent of the recirculation rate, depending on the type and efficiency of the cooling tower. Some of the more modern towers may have drift values below 0.01 percent. Learning Aide 7–1 shows a set of straightforward calculations for determining evaporation, blowdown, and makeup in a cooling tower.

Several materials of construction for a cooling tower are possible. For many years, Redwood or Douglas Fir served as the principal material for mechanical draft towers. Other materials have, to some extent, supplanted wood-frame cooling towers. These include concrete, fiberglass-reinforced-plastics, and even ceramics. Concrete is the material of choice for hyperbolic cooling towers because of the tower size and need for structural strength. Galvanized steel is often used for smaller, pre-fabricated commercial towers, but it is very susceptible to corrosion, particularly because a cooling tower constantly replenishes the oxygen content of the circulating water. Stainless steel provides much better protection, but at a much higher material cost. Its use is limited to small, commercial towers.

Closed Cooling Systems

Closed cooling systems require very little makeup and are often used for small plant heat exchange systems, such as coolers on pump bearings and building heating units. Because water losses in a closed system are minor, chemical treatment of these systems is relatively straightforward. We will look at one practical example that is an exception, however.

Cooling water systems may suffer from corrosion, scale formation, non-microbiological fouling, and microbiological fouling. Often these are interrelated. Fouling for instance may generate severe under-deposit corrosion.

Cooling Water Corrosion, Scale and Fouling Mechanisms, and Treatment Methods

Due to water's ability to dissolve, to some extent, most inorganic substances, and its ability to support microbiological life, every cooling water system is subject to potential operational problems. Scaling and fouling restrict heat transfer and generate under-deposit corrosion cells. Microorganisms generate corrodents directly through their metabolic processes. Slime produced by microorganisms increases fouling potential. Corrosion products may dislodge and end up as deposits in heat exchangers. Macroorganisms like Asiatic clams and zebra mussels will foul intake structures and heat exchangers. The source of water, type of system (once-through, recirculating, closed), and performance of treatments programs have a great impact on these phenomena.

Corrosion

Corrosion control requires careful evaluation because of the many and varied corrosion mechanisms possible. Cooling water systems may contain several different metals, including carbon-steel water lines, stainless steel or copper-alloy heat exchanger tubes, and galvanized structural supports, bolts, nuts, etc., in cooling towers. These may all suffer corrosion, sometimes from chemicals added for scale or foulant control. Concrete and wood also suffer from attack and degradation.

As we learned in chapter 1 and reviewed in chapter 4, corrosion of metals is an electrochemical process in which an electron transfer takes place between the metal and corrodents within the water. The driving force for corrosive reactions is the electrical potential between the electron acceptor (the corrosive medium) and the electron donor (the metal). Every metal exhibits a different tendency to release or accept electrons. Let's look at some of the principal corrosion mechanisms in cooling water.

Recall from chapter 1 that the baseline half-cell oxidation-reduction reaction is this:

$$2H^+ + 2e^- \rightarrow H_2 \qquad [Eq.\ 7\text{--}1]$$

This is also the half-cell equation for the reaction of acids. The reaction has, by definition, an electrical potential of 0.00. Metals like iron and zinc have lower potentials and are reactive in acid solutions, while other elements, such as copper, are passive. A number of factors influence the rate of reaction of metals in acidic waters, but the most important is the acid concentration. The higher the hydrogen ion concentration, the faster the reaction proceeds. This is the primary reason that the pH of many cooling water treatment programs is maintained within an alkaline range. The reduced hydrogen ion concentration limits reaction 7–1.

If acids were the only potential corrodent in a cooling water system, corrosion would be easier to control. But, many other corrodents exist, of which the prime culprit is oxygen. At the risk of being very redundant, I am showing again the corrosion of iron-based materials and oxygen, as is illustrated in Figure 7–4. The following processes make up the corrosion cell:

- Iron releases electrons, which travel through the metal to another site where they reduce oxygen and water to hydroxide ions. The site of reduction is known as the cathode. The local pH at the cathode increases due to the formation of hydroxide.

- Oxidized iron atoms (Fe^{+2}) leave the metal substrate and enter the solution. This site is known as the anode.

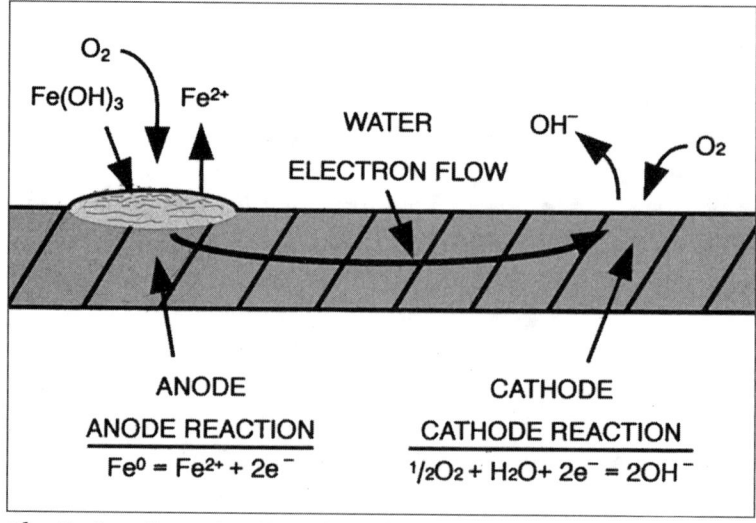

Fig. 7–4 *Corrosion Reaction of Iron by Oxygen*
Source: Betz Handbook of Industrial Water Conditioning, Ninth Edition. BetzDearborn is a division of Hercules, Inc.

- Iron and hydroxide ions combine to form a precipitate. Under the conditions shown, the precipitate eventually converts to rust ($Fe_2O_3 \cdot xH_2O$).

The corrosion cell is written thusly:

$$2Fe \rightarrow 2Fe^{+2} + 4e^- \quad \text{(Anodic reaction)} \quad [Eq.\ 7\text{--}2]$$

$$\underline{O_2 + 2H_2O + 4e^- \rightarrow 4OH^-} \quad \text{(Cathodic reaction)} \quad [Eq.\ 7\text{--}3]$$

$$2Fe + O_2 + 2H_2O \rightarrow 2Fe(OH)_2 \quad [Eq.\ 7\text{--}4]$$

$Fe(OH)_2$, ferrous hydroxide, eventually converts to rust ($Fe_2O_3 \cdot xH_2O$) by further reaction with dissolved oxygen.

Equation 7–3 is the principal cathodic reaction in neutral or alkaline solutions. Other well known cathodic reactions include:

$$O_2 + 4H^+ + 4e^- \rightarrow 2H_2O \quad [Eq.\ 7\text{--}5]$$

$$2H_2O + 2e^- \rightarrow H_2\uparrow + 2OH^- \quad [Eq.\ 7\text{--}6]$$

Several important aspects of the corrosion process should now be mentioned. First, as Equations 7–2 and 7–3 illustrate, electron flow is balanced between the anodic and cathodic half-cell reactions. Severe corrosion results when limited anodic sites are present in a large cathodic environment. Such anodic sites occur beneath deposits. The liquid underneath the deposit becomes oxygen depleted, causing the metal at the site to become anodic to bare metal locations where oxygen is in good supply. As corrosion proceeds, the pit continues to increase in depth and may eventually penetrate the tube wall. Pitting is a very insidious mechanism, and material failure can occur with relatively small metal loss. Second, corrosion via the mechanism shown in Equation 7–4 has been common in systems treated with acid for scale control. Overfeed of acid increases the hydrogen ion concentration.

Microbiologically-influenced-corrosion (MIC) is another common problem, and is particularly troublesome to iron-based materials including stainless steels. Microbes, after attaching themselves to a metal surface, secrete a carbohydrate-based gelatinous material as a protective layer. The bacteria and accompanying slime layer by themselves can generate under-deposit corrosion due to the formation of oxygen-differential cells. However, certain microbes also produce corrosive chemicals through their own metabolic processes. Among these chemicals are sulfuric acid and sulfides, both of which will attack the underlying base metal. It is very important to control microbiological fouling, because of the multiple problems that such deposits can generate. Practical Example 4-4 in a previous chapter outlined a case history of MIC in a stainless-steel condenser.

Chloride pitting of stainless steels is another potential source of concern, especially in open recirculating systems where the concentrating effect raises all levels of impurities. The two most popular stainless steels for condenser tube material are 304 and 316. Recommended maximum chloride levels for these steels range from 500 ppm for 304 SS to 3,000 ppm for 316L SS.

As was illustrated in chapter 4, copper alloys form an oxide layer when placed in service. This layer tends to protect the base metal from corrosion. Also, copper is a semi-noble element and does not undergo simple acid attack. Copper alloys can corrode, however, in the presence of compounds that complex or bond with copper. We have seen that ammonia is a primary copper complexing agent. Although the concentration of ammonia in many natural waters is too low to be a problem, if the intake is downstream from the discharge of a wastewater treatment plant, ammonia levels may reach problematic proportions.

Sulfide is another troublesome corrodent, and copper alloys should not be used to handle waters that contain sulfides. Sulfides can show up in polluted waters. Some waterside copper corrosion problems may show up as dezincification or denickelification of Admiralty metal and copper-nickel alloys, respectively. These corrosion mechanisms are influenced by excessive chlorination, low pH, and high chlorides. In these processes, the alloying element, either zinc or nickel, leaches away from the material leaving a weaker copper base.

Copper-alloys in general are more resistant to microbiological fouling than are iron-based materials. This is due to the fact that copper-alloys tend to leach Cu^{+2} ions into solution, and these ions are toxic to aquatic organisms. This resistance is partly a function of material composition and pH. More ions tend to enter solution at a lower pH. From my own personal experience, I found that Admiralty metal was more resistant to microbiological deposition than either 90-10 or 70-30 copper-nickel. Even so, the Admrialty tubes fouled at times, and especially if a biocide feed system shutdown stopped the flow of treatment chemical.

Oxygenated acid solutions are detrimental to copper alloys. I know of one utility condenser that suffered severe copper corrosion when a sulfuric acid feed system malfunctioned and lowered the pH of the oxygen-laden cooling water.

Corrosion Influencing Mechanical Factors

Corrosion is influenced by mechanical and physical factors. Anodes and cathodes will develop in a metal just due to surface irregularities. Temperature quite often affects corrosion rates. Higher temperatures usually increase the corrosion rate, just as they do for many chemical reactions. Corrosion rates of steel are generally much higher in summer than winter, as is illustrated in Figure 7–5.

Galvanic corrosion also occurs in cooling water systems, and the effects must be taken into consideration for any system containing dissimilar metals. Methods to minimize this problem include limiting the anode/cathode ratio, using impressed-current cathodic protection or sacrificial anodes, coating the metal surfaces to prevent contact with the cooling water, or electrically insulating dissimilar metals. Dielectric unions connecting copper and iron pipes are a common example of this latter method. Sacrificial anodes are a well known corrosion control method. These can be suspended in the inlet water boxes. Zinc is an often-used material, as it is more reactive than iron. The zinc anode preferentially corrodes rather than the steel waterboxes. Those interested in sacrificial anodes or cathodic protection should consult a corrosion expert before installing such devices.

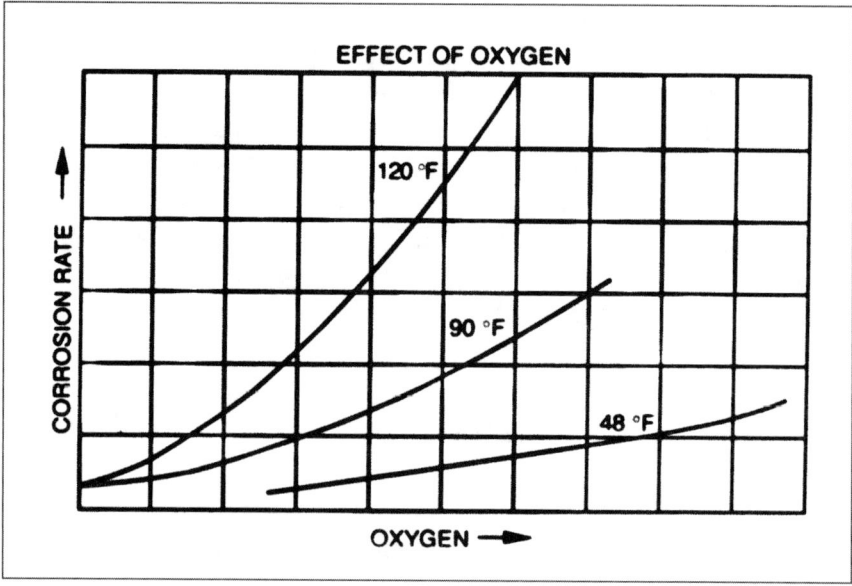

Fig. 7-5 *Effect of oxygen Corrosion on Steel as a Function of Temperature*
Source: Betz Handbook of Industrial Water Conditioning, Ninth Edition.
BetzDearborn is a division of Hercules, Inc.

Non-Metallic Corrosion

Non-metallic cooling water system materials degrade also, albeit by different mechanisms than those shown above. Wood may be subject to degradation (rot) by various types of fungi, which attack either the cellulose or lignin structure of the material. Concrete degradation may be caused by too many or too few ions in the water. Both chloride and sulfate will attack standard concrete, although special types of concrete are available to resist these ions. Sulfate attack may be more severe in cooling towers operated at high cycles of concentration, where sulfuric acid is used to control pH. Chlorides at high concentration will attack the reinforcing bars and cable in concrete.

Conversely, concrete may corrode if the water is too soft, because calcium ions will leave the concrete to establish equilibrium with the water. Supplemental calcium additions to the water may be needed to minimize this reaction. Protective coatings that serve as moisture barriers can alleviate this problem.

Corrosion Inhibitors

Corrosion inhibitors function by protecting the material surface. Since corrosion is influenced by a wide variety of conditions, it is often difficult to precisely predict when and where it will occur. A corrosion inhibitor program should be designed to provide wide-ranging protection.

Inhibitors effectively depolarize (reduce or stop the electrical flow of) the corrosion reaction either at the anode or cathode, or both when inhibitors are combined. In general, cathodic inhibitors precipitate at the locally high pH cathodic site to form a barrier that limits the rate of oxygen reduction. Anodic inhibitors typically promote the formation of a stable metal oxide at the anodic surface. This limits metal dissolution. The most common mild-steel corrosion inhibitors include:

- Anodic — Molybdate, Orthophosphate, Nitrite, Silicate
- Cathodic — Zinc, Polyphosphate, Phosphonate

The mechanisms by which the anodic inhibitors perform is interesting. Molybdate (MoO_4^-) combines with iron ions formed at anodes to produce a molecular ferric-molybdate film at the metal surface. If molybdate residuals are properly maintained, the film will eventually cover the entire metal surface. Orthophosphate (PO_4^{-3}) actually serves as both an anodic and cathodic inhibitor. It forms phosphate complexes with metal ions at the anode, but when calcium is present in the water will form a calcium phosphate precipitate at the cathode. Nitrites (NO_2^-), which are generally restricted to closed cooling water treatment, actually passivate metal and cause it to form its own protective layer. Silicates react with metal ions at the anode to form a protective film.

Anodic corrosion inhibitors can cause difficulties if not properly applied. Molybdate films are not exceptionally strong when used alone, and may require a high molybdate residual (50 ppm or more) to maintain the protective layer. This can become expensive. Orthophosphate may form calcium and iron phosphate sludges that retard heat transfer in heat exchanger tubes. Nitrites are a source of nutrition to microbes, which almost totally eliminates their use in open cooling systems. (Even in closed cooling systems, nitrite may spur the growth of microbes. See

Practical Examples 7–1 and 7–2 for discussions of nitrite applications to closed cooling systems). Silicates, which are weak inhibitors anyway, may be inappropriate for cooling towers, because silica may begin to form tenacious deposits when silica concentrations reach 150 ppm.

Underdosage of anodic inhibitors will spawn anodic sites in a large cathodic environment. This differential area effect can cause severe pitting at the anodic sites. If a proper residual cannot be maintained, it is often better to suspend the treatment program.

Practical Example 7-1

The utility has six closed cooling systems, three for building heat and three for bearing cooling water. The original chemical treatment program for all three was manual addition of sodium nitrite ($NaNO_2$) through pot feeders. The minimum control concentration was set at 2000 ppm nitrite. In all but two of the systems, air in-leakage at unknown locations would oxidize the nitrite to nitrate. This had the potential of producing limited anodic sites in a large cathodic field. Rather than risk the potential of short-term pipe failure due to pits, the utility suspended chemical treatment. In the almost two decades since, none of the systems have suffered any major leaks.

Practical Example 7-2

The following occurred in a very large closed cooling water system at a major automotive assembly plant. The system cools the tips of automatic welders. At the time of the incident, sodium nitrite was the primary treatment chemical. Plant personnel noticed that many of the welders were overheating. When they opened the system, extensive microbiological deposition was evident. Nitrifying bacteria had been feeding off the nitrite. The plant switched to molybdate and the bacterial growth stopped.

Cathodic inhibitors often perform similarly to anodic inhibitors with regard to protective deposit formation. Zinc combines with hydroxide to form a precipitate at the cathode. Polyphosphates, like orthophosphate, combine with calcium and deposit at the cathode. Phosphonates (organo-phosphorous compounds) also form a deposit on the cathodic surfaces.

A benefit of cathodic inhibitors is that if underdosed, they do not cause the formation of anodic sites in a large cathodic field. However, cathodic inhibitors are not foolproof. Zinc salts are not particularly tenacious and may wash away. Zinc is also toxic to some aquatic organisms, and its use has been curtailed due to this factor. Polyphosphates break down to form orthophosphate, which can increase scale and sludge formation in the system. In fact, calcium phosphate now rivals calcium carbonate as the most common scale, on average, in cooling water systems. Phosphonates are degraded by oxidizing biocides, which reduces the effectiveness of the inhibitor and generates orthophosphate.

As these previous examples should indicate, a single inhibitor is often inadequate, and a combination of inhibitors may be more effective. For example, molybdate and zinc will provide anodic and cathodic protection. Molybdate and orthophosphate provide combined anodic protection, and a zinc/phosphonate program provides combined cathodic protection. Table 7–1 illustrates recommended dosage ranges for various combined inhibitor programs.

Two factors have had a great impact on choice of corrosion inhibitors. One has its basis in environmental issues and the other is the general trend towards alkaline cooling water treatments.

Common Corrosion Inhibitors and Dosage Levels

Inhibitor	Dosage Range (ppm)
Molybdate	25–50
Orthophosphate	5–15
Polyphosphate	1–30
Phosphonates	5–10
Molybdate/Phosphate	5–10/5–10
Molybdate/Phosphonate	5–10/5–10
Zinc/Phosphate	1–3/5–10
Zinc/Phosphonates	1–3/5–10

Table 7–1 *Recommended Dosage Ranges for Corrosion Inhibitor Programs*

Environmental regulations governing cooling water discharges are becoming more and more strict. Specifically, metal concentrations in cooling tower blowdown directly discharged to a receiving body of water are being severely curtailed. Chromate, which was once the most popular and effective corrosion treatment chemical, has almost totally disappeared. The new focus is directed towards zinc. Limits of 1 ppm in cooling water discharge streams are common. Molybdenum has not come under such attack yet, because it is not very toxic to aquatic life.

Alkaline cooling water treatments, which tend to minimize direct corrosion, have become very popular. This popularity is in part due to the restrictions on chromate. Where corrosion inhibitors are still used, the trend is toward phosphate/phosphonate/polymer treatments. The phosphate and phosphonates combine with calcium to produce deposits at the cathode, but the polymer modifies the deposit structure to prevent heavy buildups of calcium phosphate. These programs will reportedly reduce mild steel corrosion below 5 mils per year (mpy), and sometimes 1 mpy.

Copper corrosion prevention has not been neglected either. These programs are commonly referred to as yellow-metal treatments, and are based upon azole chemistry. Two common azoles, tolyltriazole (TTA) and butylbenzotriazole, are shown in Figures 7–6 and 7–7, respectively. The azoles function by bonding to the copper metal through nitrogen atoms on the molecule. The flat azole molecules form a barrier on the metal

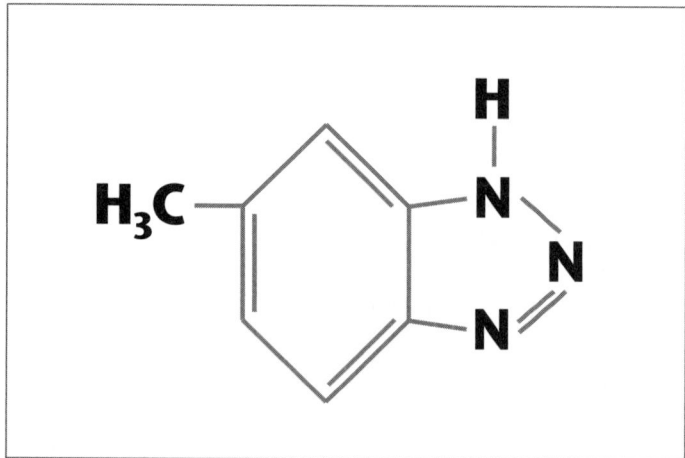

Fig. 7–6   *Tolyltriazole*

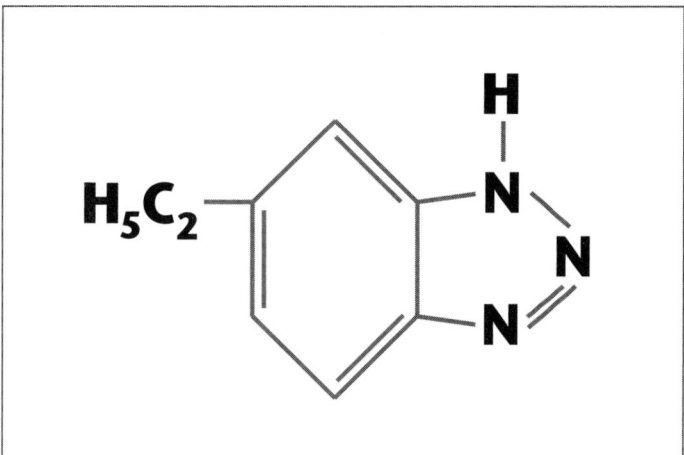

Fig. 7–7   *Bulylbenzotriazole*

surface. Azole dosages of 1 to 2 ppm are common. The azoles, which function more efficiently in basic solutions, have become popular as an additive to alkaline scale inhibitors. Oxidizing biocides will degrade azoles, so feed programs must be designed to take this into account. However, at least one new azole, which is resistant to degradation by oxidizing biocides, has appeared on the market.

Scale

Scale and deposit control are very important in cooling systems. Scale buildups on heat exchanger tubes and cooling tower film fill inhibit heat transfer and flow patterns. Even worse, scale buildups can initiate corrosion reactions, where corrosion occurs underneath the deposits. Failures may occur even though most of the metal is still in decent condition. We will examine the most common scales and popular treatment methods. Scale formation is of much greater concern than before because of the movement away from acid/chromate programs to alkaline treatments.

Calcium Carbonate Scale

Due to the abundance of calcium and bicarbonate alkalinity in many natural water supplies, calcium carbonate has led the list of most frequently encountered scales in cooling systems. Its place has, however, been threatened by calcium phosphate because phosphates and phosphonates have become such popular treatment compounds. The most common cooling water scales are

- Calcium phosphate
- Calcium carbonate
- Silica
- Calcium sulfate
- Calcium fluoride

The potential for scale formation is much higher in open recirculating cooling systems, where the dissolved solids concentration may be 5, 10, or perhaps 20 times greater than that in the makeup. Calcium carbonate scale forms when calcium and alkalinity exceed saturation values and begin to drop out as calcium carbonate ($CaCO_3$). This reaction is strongly influenced by temperature, and waters that may not exhibit scaling tendencies at ambient temperatures will cause big problems inside a condenser where water temperatures may rise 30° or more.

A number of programs are available to estimate scaling potentials. In the 1930s, Langelier developed the first set of calculations for this purpose, and the value derived from the calculations is known as the Langelier Saturation Index (LSI). The LSI is still used today. Calcium carbonate precipitation is dependent not only on the calcium and alkalinity in solution but also on the pH, and, of course, temperature. Langelier's equations take this into account:

$$pH_s = (pK_2 - pK_s) + pC_a + pAlk \qquad [Eq.\ 7\text{–}7]$$

$$LSI = pH_a - pH_s \qquad [Eq.\ 7\text{–}8]$$

CHAPTER 7: COOLING WATER CHEMISTRY

where

pH_s is the Langelier Saturation pH

pK_2 is the second dissociation constant for carbonic acid

pK_s is the solubility product constant for calcium carbonate

pC_a is the negative log of the calcium concentration

$pAlk$ is the negative log of the alkalinity concentration as based on a titration procedure

pH_a is the actual pH

An LSI of less than zero indicates that the water tends to dissolve scale, and thus may be corrosive. An LSI greater than zero indicates that the water may be scale forming. Nomographs have been developed for calculating the LSI. One of these is shown in Figure 7–8. The user determines the values for pCa, $pAlk$, and C, whose sum is then subtracted from the actual pH

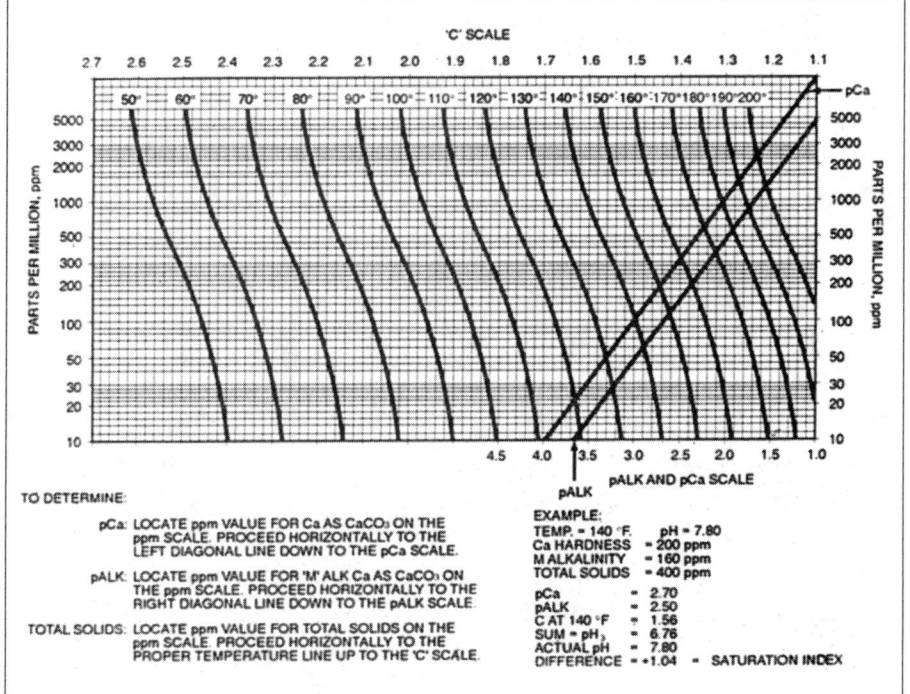

Fig. 7–8 *Nomograph to Calculate LSI*

along the correct temperature line. This gives the LSI. The LSI can also be approximated from the following equation:

$$LSI = pH - [9.4 - \log_{10} Ca - 0.97 \log_{10} M_{ALK} + (\log_{10} TDS/10.7) - 3.24 \exp(-T/191)] \quad [Eq.\ 7\text{--}9]$$

where

Ca is calcium as mg/l, $CaCO_3$

M_{ALK} is total alkalinity as mg/l, $CaCO_3$

TDS is total dissolved solids as mg/l

T is the heat exchanger water-side skin temperature, °F

Although the Langelier Saturation Index has proven useful, it does not always provide the accuracy needed due to inherent errors in the calculations. In particular, the calculations do not account for ionic interactions in the water. To improve the accuracy of the calculation, Ryznar developed the Stability Index:

$$RSI = 2pH_s - pH_a \quad [Eq.\ 7\text{--}10]$$

The scale of the RSI is different than that of the LSI. Waters with an RSI of less than 6 are scale-forming, while those above 7.5 exhibit corrosive tendencies. The LSI and RSI are often used in combination, although the RSI is based on the same calculations as the LSI and provides no additional improvement with regard to ionic interactions.

Paul Puckorius has developed his Practical Scale Index (PSI), which incorporates temperature and conductivity into the calculations, and more explicitly takes into account the effect of alkalinity on the chemistry of the solution. The PSI calculation, whose scale is similar to the Ryznar index, is available in slide-rule form. A water with a PSI of greater than 6.0 is considered to be scale-dissolving or corrosive, and less than 6.0 to be scale-forming. Table 7–2 shows a side-by-side comparison of the indices. Even though the RSI and PSI scale are the same, the values obtained from the two are often not equal, because the PSI calculates a pH equilibrium value (pH_e), which is often different than pH_a.

Water Cooling Scale

LSI	RSI	PSI	Condition Without Treatment
3.0	3.0	3.0	Extremely severe scaling
2.0	4.0	4.0	Very severe scaling
1.0	5.0	5.0	Severe scaling
0.5	5.5	5.5	Moderate scaling
0.2	5.8	5.8	Slight scaling
0.0	6.0	6.0	Stable water
–0.2	6.5	6.5	Very slight scale dissolving
–0.5	7.0	7.0	Slightly scale dissolving
–1.0	8.0	8.0	Moderately scale dissolving
–2.0	9.0	9.0	Strongly scale dissolving
–3.0	10.0	10.0	Very strongly scale dissolving

Table 7–2 *Comparison of Water Cooling Scale Indices*

The indices may lose applicability for analyzing waters treated with calcium carbonate crystal modifiers, because calcium carbonate formation is inhibited. However, the programs are still useful for determining the scale or corrosion tendencies of makeup water, and are helpful for designing new systems.

More accurate programs are now available for determining calcium carbonate scale potential. The personal computer has made these possible. Some water treatment software companies offer scale-calculation computer programs for sale. The major water treatment chemical firms also have predictive programs that can readily calculate the scaling potential of cooling waters.

Other Scales

Many other scales are possible in cooling water systems. Calcium phosphate is obviously on the list due to the increased popularity of phosphate and phosphonate treatment programs. Calcium sulfate is still there, although the declining use of acid treatment has, in general, made this less common. Calcium sulfate is a tough scale to remove. Silica scale can be even worse, because silica deposits are very tenacious. As we learned in our examination of scale formation in boilers, silica may precipitate by itself or in combination with a number of other ions. In cooling water systems, magnesium silicate is particularly troublesome.

The following are reported guidelines for maximum levels of calcium and other anions besides bicarbonate that can be tolerated without treatment.

Calcium Phosphate. Calcium is limited by the following equation when phosphate concentrations are 10 ppm or higher.

$$Ca\ (as\ ppm\ CaCO_3) < (105 \times (9.8 - pH))$$

Calcium Sulfate

$$Ca\ (ppm\ as\ CaCO_3) \times SO_4^{-2} < 1,800,000\ ppm$$

Silica. Silica concentrations are maintained below 150 ppm to prevent formation of silica deposits. The limit can sometimes be extended to 200 ppm or higher with proper dispersants.

Magnesium Silicate

$Mg\ (ppm\ as\ CaCO_3) \times SiO_2 < 500,000$	at pH of 7
$Mg\ (ppm\ as\ CaCO_3) \times SiO_2 < 100,000$	at pH of 7.5
$Mg\ (ppm\ as\ CaCO_3) \times SiO_2 < 70,000$	at pH of 8
$Mg\ (ppm\ as\ CaCO_3) \times SiO_2 < 50,000$	at pH of 8.5
$Mg\ (ppm\ as\ CaCO_3) \times SiO_2 < 10,000$	at pH of 9

The reader should note that these are only general recommended guidelines, and as with all other guidelines in this book, each system must be evaluated individually.

Please note the magnesium silicate guidelines. The allowable concentration greatly decreases as pH increases. Unfortunately, this conflicts with the trend towards alkaline cooling treatment programs and/or higher cycles of concentration.

Scale Control

A number of scale control techniques and chemical programs are available. These range from traditional programs such as acid feed, to treatment with complex organic polymers. The range of chemicals gives the operator flexibility in program selection.

Scale Control by Acid Feed

For many years, the primary method for controlling calcium carbonate scale was sulfuric acid addition. Sulfuric acid reduces bicarbonate alkalinity as follows:

$$H_2SO_4 + Ca(HCO_3)_2 \rightarrow CaSO_4 + 2CO_2\uparrow + 2H_2O \quad [Eq.\ 7\text{--}11]$$

Although this treatment method is not as common as before, it still can be useful for pretreating waters high in hardness and alkalinity. Before discussing acid feed rates, let's re-examine alkalinity and its relationship in water.

Carbon dioxide, bicarbonate, and carbonate exist in equilibrium in solution, with the amount of each dependent upon pH. The equilibrium reactions are the following:

$$H_2O + CO_2 \leftrightarrows H_2CO_3 \leftrightarrows HCO_3^- + H^+ \quad [Eq.\ 7\text{--}12]$$

$$HCO_3^- \leftrightarrows CO_3^{--} + H^+ \quad [Eq.\ 7\text{--}13]$$

At a pH of less than 4.4, only free carbon dioxide is present. Between a pH of 4.3 and 8.3, carbon dioxide is replaced by the bicarbonate ion until at pH of 8.3 essentially no free carbon dioxide remains. Between pH 8.3 and 12, bicarbonate gradually converts to carbonate. At a pH of 12 or above, hydroxide is present. Although pH can be used to measure the relative basicity of a solution, the complexity and buffering effects of the carbon dioxide-bicarbonate-carbonate series require additional measurements for precise calculation. Bicarbonate, carbonate, and hydroxide (for lime softened waters) levels are determined by a series of titrations. Titration of an alkaline solution with acid to a pH of 4.3 using a methyl orange indicator determines all of the alkalinity in solution. This is known as the M alkalinity. Solutions that contain carbonate ion may be titrated with phenolphthalein as an indicator. The phenolphthalein changes color from pink to clear at a pH of 8.3. This titration determines all of the hydroxide and half of the carbonate ion in solution. If one performs both a P and M titration on a sample, the amount of bicarbonate, carbonate, and hydroxide may be determined. Table 7–3 illustrates the calculations for these determinations.

HCO_3, CO_3, and OH Relationship to P and M Readings

Alkalinity Reading (ppm as $CaCO_3$)	Hydroxide Conc. (ppm as $CaCO_3$)	Carbonate Conc. (ppm as $CaCO_3$)	Bicarbonate Conc. (ppm as $CaCO_3$)
P=0	0	0	M
P=M	P	0	0
2P=M	0	2P	0
2P<M	0	2P	M–2P
2P>M	2P–M	2(M–P)	0

Table 7–3 *Relationships of HCO_3, CO_3, and OH to P and M Analyses*

Acid feed to a cooling tower is predicated on maintaining alkalinity levels below those which would cause calcium carbonate precipitation. It may be tempting to closely monitor the alkalinity of a solution and control the acid feed so that it always just keeps the pH low enough to prevent calcium carbonate formation. This can be perilous, because changes in temperature and makeup water quality, failure of the acid feed pump, or poor calibration of a pH monitor could all cause conditions to change to scale-forming. The upset might not be noticed for some time, further exacerbating the situation. It is more common to establish a safety margin between the actual chemistry and the point where scaling would occur. By keeping the M alkalinity of the recirculating water at approximately 40 ppm, the pH can be controlled in a neutral range near 7.0. This precludes the formation of carbonate ions, and maintains a buffer below the pH (8.3) at which carbonates would form.

Two major problem with sulfuric acid treatment is that it lowers the pH and increases the corrosive tendencies of the cooling water, and it increases calcium sulfate scaling tendencies. Upsets in feed rate magnify these problems. Cases are known where a malfunctioning feed system has caused excess acid introduction, severe pH depression, and corrosion of cooling system components. This is yet another factor in the trend towards alkaline cooling water treatment.

Alkaline Treatment Methods

In a few cases, utilities have resorted to pretreatment by clarification and softening of cooling water to reduce hardness. This is usually not economical unless the water is very hard. More popular are treatments that function in-situ. In an alkaline program, the pH of the cooling water is allowed to equilibrate at a natural level, or may even be pushed into a basic pH range by addition of alkaline compounds such as caustic soda, potassium hydroxide, or soda ash. Naturally, this increases the potential for calcium carbonate scale formation. Therefore, alkaline treatment programs are designed to either keep calcium in solution or to modify the crystalline structure of calcium precipitates so that they form a sludge-like product, which can be blown down.

Solublizing compounds include phosphonates and polymers such as polyacrylate. Let's briefly look at this chemistry. One of the most popular of the original group of phosphonates is hydroxyethylidene diphosphonic acid (HEDP, Figure 7–9). The phosphate portion of this compound adsorbs onto newly-formed calcium carbonate crystals, distorting crystal shape and slowing growth. An advantage of phosphonates over phosphate is that the compounds do not by themselves form calcium orthophosphate, which causes sludge buildups in cooling systems.

Phosphonates may produce some negative side effects. They are corrosive to zinc, and mildly corrosive to copper and aluminum. Phosphonates also break down in the presence of oxidizing biocides, par-

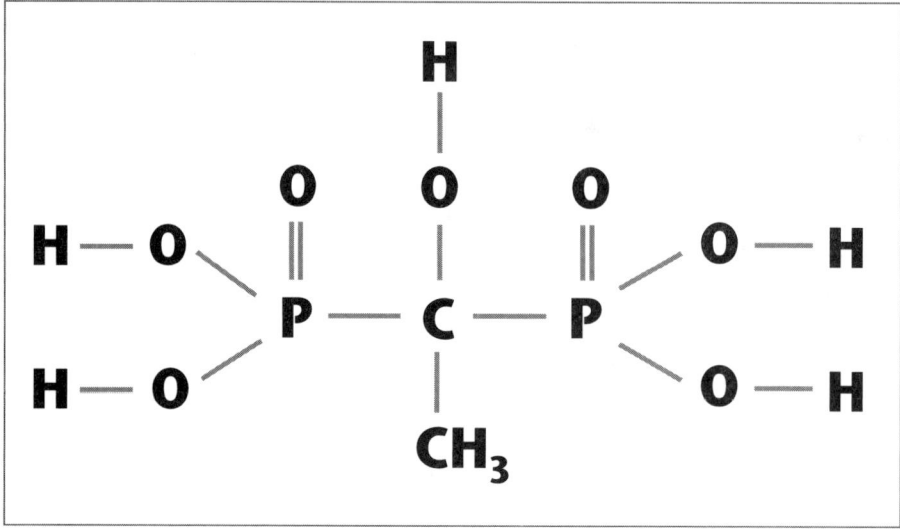

Fig. 7-9 *Hydroxyethylidene Diphosphonic Acid*

ticularly at biocide concentrations above 1 ppm. Not only will this cause the phosphonates to lose effectiveness, but the breakdown products include orthophosphate, which can combine with calcium to form calcium phosphate scale. Despite these factors, phosphonates are still the most popular scale control agents. Aminoethylenephosphonic acid (AMP) and a newer product, phosphono-butane-tricarboxylate (PBTC) are alternatives to HEDP. PBTC in particular has made an impact. It is reported to be less corrosive than other phosphates to copper, and more resistant to degradation from oxidizing biocides or UV light.

Polyacrylates (Figure 7–10) are one of the popular blends in a phosphonate/polymer program. They also keep calcium in suspension. The polymer binds calcium through the partial negative charges on the oxygen atoms. A co-polymer of polyacrylate and acrylamido-2-methylpropane-sulfonic acid has proven to be a very effective dispersant.

Crystal modifiers include polymaleates and co-polymers of these with sulfonated polystyrenes. These polymers allow calcium deposits to form, but alter the structure so that the deposits do not precipitate as a hard scale. The solids exit the system in the blowdown or sidestream filter. Crystal modifiers are more expensive than other treatments and are not used nearly as often.

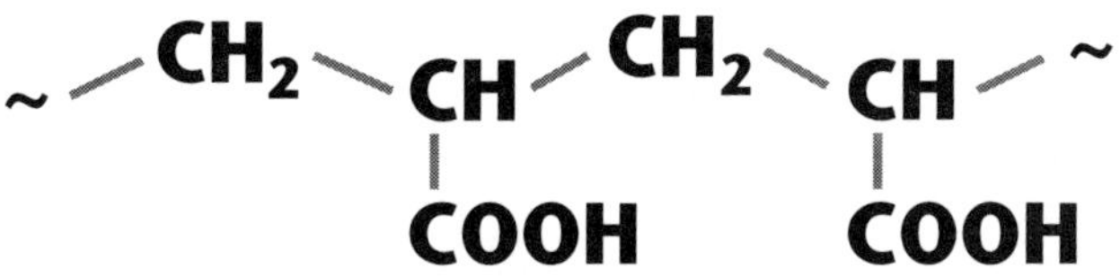

CHAPTER 7: COOLING WATER CHEMISTRY

$$\sim\!\!-\!\!CH_2\!\!-\!\!CH\!\!-\!\!CH_2\!\!-\!\!CH\!\!-\!\!\sim$$
$$\qquad\quad |\qquad\qquad\quad |$$
$$\qquad\,COOH\qquad\,COOH$$

Fig. 7–10 *Polyacrylate*

Table 7–4 illustrates guidelines for scale inhibitor residuals. With all of these options, it is often difficult to decide on a treatment program. However, some trends have emerged. Blended products of phosphonates and polymers seem very popular. One common blend contains HEDP and PBTC for calcium carbonate control (HEDP also provides corrosion inhibition for steel), and a polymer for calcium phosphate control. Azoles may be added for copper corrosion control.

Common Scale Inhibitors and Dosage Levels

Inhibitor	Dosage Range (ppm)
Phosphonate	3–5
Polyphosphate	3–5
Polymaleics	1–2
Sulfated Polystyrene	1–2

Table 7–4 *Guidelines for Scale Inhibitor Residuals*

Phosphate/phosphonate treatments with polymer scale dispersants still remain in use, although one major water treatment company has developed a new, short-chain polymer that contains no phosphorous. The compound, formed by polymerization of an epoxide, contains many oxygen atoms along the polymer backbone for inhibiting calcium carbonate scale. The polymer is reported to be very popular and has allowed the company to substitute it in place of phosphate-phosphonate compounds in a number of applications.

With these programs, or various combinations thereof, it is reported that calcium, sulfate, and phosphate concentrations may be allowed to reach the following limits:

Calcium Phosphate

    Ca (as $CaCO_3$) — 1200 ppm

    PO_4 — 30 ppm

Calcium Sulfate

    Ca (as $CaCO_3$) — 2700 ppm

    SO_4 — 4000 ppm

LSIs as high as 2.8 to 3.0 may be possible with the correct chemical program. It has also been claimed that these new polymer formulations allow higher silica concentrations of up to 300 ppm. Some water treatment experts view this cautiously, and still abide by the old silica limit of 150 ppm.

The selection of the best cooling water treatment is dependent upon water quality, process conditions, and system metallurgy. The choice is sometimes not easy. Any program should be monitored closely to determine its performance.

Fouling

Fouling is the deposition of suspended solids or buildup of microbiological organisms within heat exchangers and cooling tower fill. Like scale, foulants reduce heat transfer, restrict water flow, and generate under-deposit corrosion cells. Foulants are often worse than scale with regard to the latter two problems.

Foulants enter cooling water systems from a variety of sources. A make-up supply from a lake or river may contain silt and debris that is stirred up during seasonal changes or heavy rainfall. Surface water sources contain many microorganisms that will cause microbiological buildups even in once-through cooling systems. A cooling tower is an ideal source for foulant introduction, because the tower is an efficient air scrubber. Warm water, aeration, nutrients, and sunlight transform a cooling tower into a great bio-reactor. The following sections discuss fouling control methods.

Non-Microbiological Fouling Control Methods

Both chemical and mechanical methods may be used to minimize fouling. One practical mechanical method is sidestream filtration of recirculating water. Sidestream filters remove silt particles, which would otherwise concentrate in the tower and cooling system. Only a relatively small percentage (1% to 5% of the recirculating flow rate) must be filtered, because the entire volume of recirculating water will pass through the filter several times every day. For example, consider a 100,000 gpm recirculating system. Assume total system volume is 300,000 gallons. A sidestream softener operating at 3% of the recirculating capacity will filter 3,000 gpm. This equates to 4,320,000 million gallons per day, or 14.4 system volumes daily. Learning Aide 7–2 outlines the calculations needed to size a sidestream filter.

Sidestream filters may be of the multi-media or cartridge filter type. One additional benefit of sidestream filters is that they will also remove iron oxide particles that may have been generated by corrosion of the cooling water lines.

Another on-line mechanical cleaning method, for condenser tubes, involves the use of sponge balls or spherical scraper-like devices that are introduced at the condenser inlet and collected by a grate system at the condenser outlet. Ball size is equal to or slightly larger than the diameter of the condenser tubes so that they will scrub the entire tube surface as they pass through. These devices are excellent for removing soft deposits like biological slime. Given the increasingly stringent restrictions being placed on chlorine and oxidizing biocides, such mechanical cleaning systems merit consideration. However, some difficulties are inherent with on-line cleaning systems. The most obvious problem is retrofitting a system to an existing condenser, for this involves installation of a collection device on the condenser outlet and recirculation line to the condenser inlet. Capital costs may be prohibitive. Operating difficulties are also possible. A malfunctioning collection system will allow the balls to escape. At least one facility reported that their scale inhibitor treatment program caused the balls to clump into sticky masses, although this could have been caused by overfeed of the treatment chemicals.

A common method for cleaning condenser tubes off-line is mechanical scraping. Maintenance personnel insert mechanical scrapers into the tubes, and then push the scrapers through with water or compressed air. The scrapers are equipped with either metallic or plastic fins to dislodge deposits. Although this procedure may take several days, it does provide the desired results. I personally have had the opportunity to work on a number of these projects, most often for removal of microbiological deposits but also for calcium carbonate scale. Tube scraping always returned the condenser to near-optimum performance. The advantage of mechanical scraping versus on-line cleaning is much lower capital cost. However, mechanical scraping requires an outage. Metallic scrapers may remove metal during their passage through the tubes.

While mechanical methods can be effective for control of foulants, their prohibitive cost or on-line limitations may make them impractical. Chemical treatment is sometimes the best or only option.

Foulant control chemicals are often similar or identical to scale control compounds. The most notable foulant treatments include low molecular weight polyacrylates, polyacrylamide, polymaleic-anhydride, and certain co-polymers. These dispersants keep foulants in suspension by reinforcing their negative surface charge, causing them to repel each other.

Table 7–5 lists typical dosages for foulant treatments. Foulant control, along with scale control, is becoming more important due to the increasing use of high-efficiency, close-spaced film fill in cooling towers.

Common Fouling Inhibitors and Dosage Levels

Inhibitor	Dosage Range (ppm)
Polyacrylate	4–5
Polyacrylamide	0.2–0.5

Table 7–5 *Typical Dosages for Foulant Treatments*

Microbiological Fouling

Microbiological fouling may be the worst problem in cooling water systems. Microbes can do the following:

- Generate acids and corrodents that attack the underlying-tube metal.

- Secrete a protective gelatinous layer that coats the tube surface. The secretions, along with silt that becomes trapped within, reduce heat transfer and cause under-deposit corrosion.

- Be pathogenic. In the most often-cited case, 34 American Legion members died in 1976 of complications caused by the bacteria Legionella pneumophilia, which was present in a hotel cooling system. My father was staying at that hotel during that exact time and came down with what he thought was the flu. Fortunately, he was in good health and did not suffer worse consequences.

Three basic types of microorganisms may exist in a cooling water system, algae, bacteria, and fungi. Algae contain chlorophyll and require sunlight to thrive. These organisms proliferate in wetted areas of a cooling tower exposed to sunlight. Fungi, including mold and yeasts, are non-photosynthetic organisms, which do not need sunlight for survival. They are most notable for attacking either the cellulose or lignin components of wood, where they induce "rot." Bacteria also do not need sunlight for survival, and may thrive in many areas, particularly in condensers where the water is warm. Bacterial counts of greater than 10,000,000 per milliliter are possible in systems with insufficient biocide feed.

Control of Microbiological Organisms

With the exception of condenser tube cleaning, it is virtually impossible to mechanically control microbiological organisms in large cooling systems. Therefore, chemical feed of a biocide is required. (Ultraviolet light and site-generated ozone have been successfully used for small cooling tower and makeup system disinfection.) Chemical treatment can be split into two basic categories, oxidizing (chlorine, bromine, chlorine dioxide) and non-oxidizing. Oxidizing chemicals attack cell components to kill the organism, while non-oxidizing biocides may damage the cell wall or interfere with the cell's metabolic processes.

Environmental and safety restrictions on the use of biocides are causing significant changes in biocide treatment. While the following sections provide a discussion on the properties and use of the most common biocides, they are also intended to give the reader an idea of the current trends in microbiological control.

Chlorine

For many years, chlorine was the primary disinfectant for all types of water systems. Its use has fallen into some disfavor because of safety issues related to gaseous chlorine, and because chlorine is known to react with organics to produce halogenated organic compounds.

Chlorine gas is the least expensive of all oxidizing biocides. When chlorine is added to water, it reacts as follows:

$$Cl_2 + H_2O \leftrightarrows HOCl + HCl \qquad [Eq.\ 7\text{-}14]$$

HOCl is the active ingredient of this mixture, and it attacks organisms very quickly. Researchers have calculated that the kill rate of a 1.0 ppm chlorine solution is 99% in 30 seconds at a pH of 6.5. The key to chlorine effectiveness is pH. The stability of HOCl is dependent upon pH, as the following reaction shows:

$$HOCl \leftrightarrows H^+ + OCl^- \qquad [Eq.\ 7\text{-}15]$$

Remember LeChâtelier's Principle from Chapter 1. This is an example. An increase in pH means an increase in hydroxyl ions (OH^-). These react with hydrogen ions to form water, so increasing pH forces the reaction to the right. OCl^- is much less potent than HOCl. Some scientists theorize that OCl^-, being a charged ion, has much more difficulty penetrating the cell wall and attacking cell components. The dissociation of HOCl begins at a pH of about 5.2, reaches 50 percent at a pH of 7.5, and is fully complete at a pH of 9.4. Chlorine's effectiveness becomes limited in an alkaline cooling water treatment program, especially as pH rises.

CHAPTER 7: COOLING WATER CHEMISTRY

Application of Chlorine

Chlorine gas is very toxic and must be handled with great care. A common method of introducing chlorine gas into water is through an eductor system, in which a sidestream of flowing water is used to pull the chlorine gas from the chlorine cylinder into the stream. Chlorine is commonly shipped in one-ton cylinders, which can be loaded in multiple units into a rack containing a manifold system for quick transfer between full and empty containers. Chlorine systems are now required to have ambient air monitoring and alarm systems that provide safety warnings in the event of a chlorine leak. The maximum allowable ambient air chlorine limit as established by OSHA is 1.0 mg/l.

Chlorine reacts irreversibly with a number of constituents in water, most notably ammonia and organics. These reactions, which reduce the amount of free chlorine available as a biocide, are known as the chlorine demand. The higher the chlorine demand, the more chlorine that must be added to achieve the same level of killing effectiveness. Ideally, a 0.1 to 0.5 ppm free chlorine residual is most effective in controlling microorganisms, while minimizing degradation of cooling water treatment chemicals or cooling tower materials.

Environmental aspects of chlorine usage are becoming more complicated. A number of years ago the EPA established "technology-based limits" for power plants that restricted the chlorine residual in the cooling system discharge to 0.2 ppm for a maximum of two hours per day. These limits are gradually giving way to "ambient water quality standards" to be determined in the stream at the boundary of a calculated mixing zone. The new limit is 0.019 mg/l in fresh water for intermittent applications. Individual states may set more stringent limits, and often apply the limits at "end-of-pipe" instead of allowing for dilution in the mixing zone.

Such concentrations are much too low to be effective. A method to maintain higher free chlorine residuals is to dehalogenate the discharge with sodium sulfite (Na_2SO_3), sodium bisulfite ($NaHSO_3$), or sulfur dioxide gas. These react with chlorine in a 1 to 1 stoichiometric fashion. An example is outlined below.

$$Cl_2 + Na_2SO_3 + H_2O \rightarrow Na_2SO_4 + 2HCl \qquad [Eq.\ 7\text{--}16]$$

An excess of sulfite is added to drive the reaction to completion. Continuous halogenation of cooling water influent with subsequent dehalogenation of the effluent before discharge is a popular macrofouling control technique, particularly at utilities and industries that have to deal with zebra mussels. This is discussed in more detail later.

There is no question that chlorine has saved countless lives from water-borne diseases since its introduction as a disinfectant. But, with the development of other disinfectants, further restrictions on chlorine use for water treatment may be in the offing. Some environmental groups are calling for a total ban on chlorine due to health concerns. With regard to cooling water treatment specifically, chlorine use is being questioned because of the halogenated organics issue.

Liquid and Solid Chlorine Donors

Where chlorine is still the preferred biocide but safety is of concern, liquid sodium hypochlorite (NaOCl) is an alternative. Sodium hypochlorite is more expensive than gaseous chlorine, but is much less hazardous to handle, can be fed directly to the cooling water, and can be stored in a bulk tank. The percentage of sodium hypochlorite in bulk industrial solutions ranges from about 10 to 15 percent. Sodium hypochlorite will decompose with time, increasing temperature, or in the presence of impurities, particularly iron and copper. Therefore, it is important to design and specify a system that minimizes hypochlorite decomposition. Learning Aide 7–3 outlines a list of specifications and guidelines for ordering and storing sodium hypochlorite.

Sodium hypochlorite undergoes the same reactions as chlorine does in water, and can produce the same compounds, including halogenated organics. It, along with solid chlorine compounds, are nowadays viewed somewhat less favorably because of this aspect. Several solid chlorine compounds are available. Calcium hypochlorite [$Ca(OCl)_2$, the common swimming pool biocide] is one. More popular for the industrial water treatment industry are the dichlorohydantions (Figure 7–11) and the chlorinated isocyanuarates (Figure 7–12). The hydantoins may be manufactured into powder, granules, pellets, or tablets, which can then be loaded

CHAPTER 7: COOLING WATER CHEMISTRY

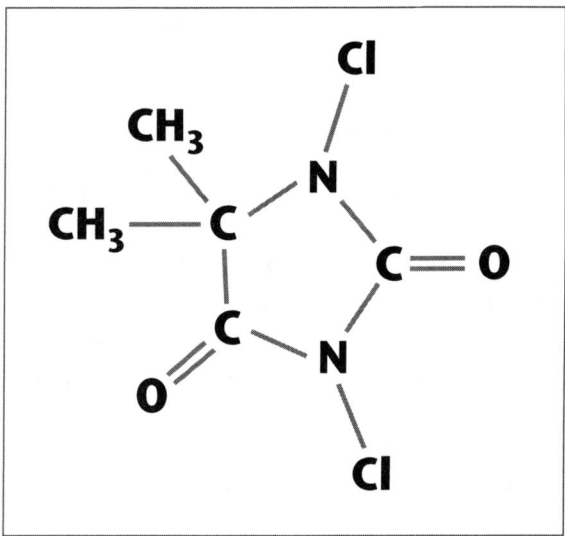

Fig. 7–11 *Dichlorohydrantoin*

into a dissolving vessel that takes a side stream of water from the cooling loop. The products are designed to dissolve at a relatively slow, constant rate to release chlorine gradually. The isocyanurates provide an added benefit by helping to keep chlorine in solution. However, the buildup of iso-

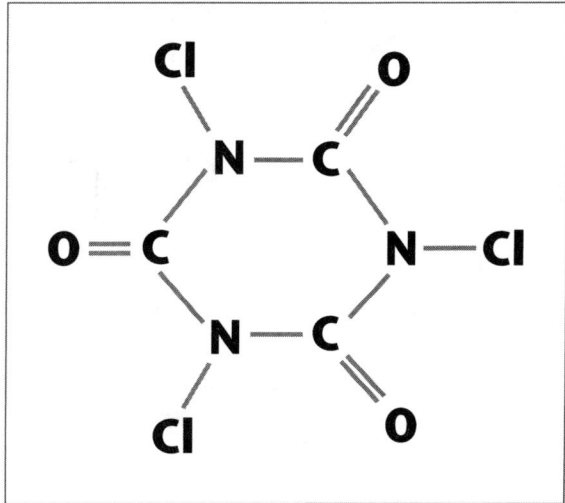

Fig. 7–12 *Chloronated Isocyanuarate*

cyanuric acid in recirculating systems can eventually limit the effectiveness of the biocide.

Chlorine Alternatives

The proliferation of alkaline cooling water treatment programs and the issue of chlorinated organics has caused a steady decline in the use of chlorine and an increase in alternate treatment methods.

A popular alternative is bromine. Bromine, which is also a halogen, is not introduced directly to a cooling water system, but rather is generated on-site by reacting a bromide solution with chlorine in the following step-wise procedure:

$$Cl_2 + H_2O \leftrightharpoons HOCl + HCl \qquad [Eq.\ 7\text{--}17]$$

$$HOCl + NaBr \rightarrow NaCl + HOBr \qquad [Eq.\ 7\text{--}18]$$

Bromine can also be generated by reaction of sodium bromide with hypochlorite.

Bromine offers several advantages over chlorine, particularly in alkaline waters. First, as Figure 7–13 illustrates, the dissociation of HOBr to OBr⁻

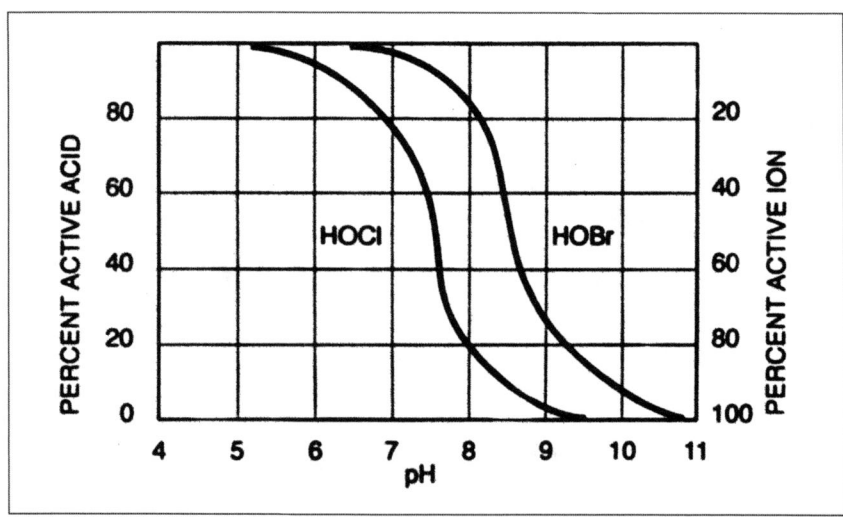

Fig. 7–13 *Dissociation of HOBr into OBr^- and H^+*
Source: Betz Handbook of Industrial Water Conditioning, Ninth Edition. BetzDearborn is a division of Hercules, Inc.

and H^+ takes place at a higher pH. Thus, bromine's biocidal efficiency in basic solution is greater. Secondly, although bromine reacts with ammonia and amines, the reaction is reversible. Unlike chlorine, bromine introduced to ammonia-laden waters still exhibits free-halogen properties.

Because bromine is similar to chlorine in biocidal efficiency, and better in alkaline waters, residual concentrations are maintained at similar levels (0.1 to 0.4 ppm).

Bromine may be generated with either chlorine gas or liquid hypochlorite. This is usually carried out with an eductor system that draws both the chlorine and sodium bromide into a side stream where the reaction takes place. Many of the major water treatment firms have developed bromine generators, some of which are very simple in design. Bromine is also supplied in solid form as hydantoins (Figure 7–14), which are introduced to cooling water in the same fashion as the chlorohydantoins mentioned above. Practical Example 7–3 discusses one such system.

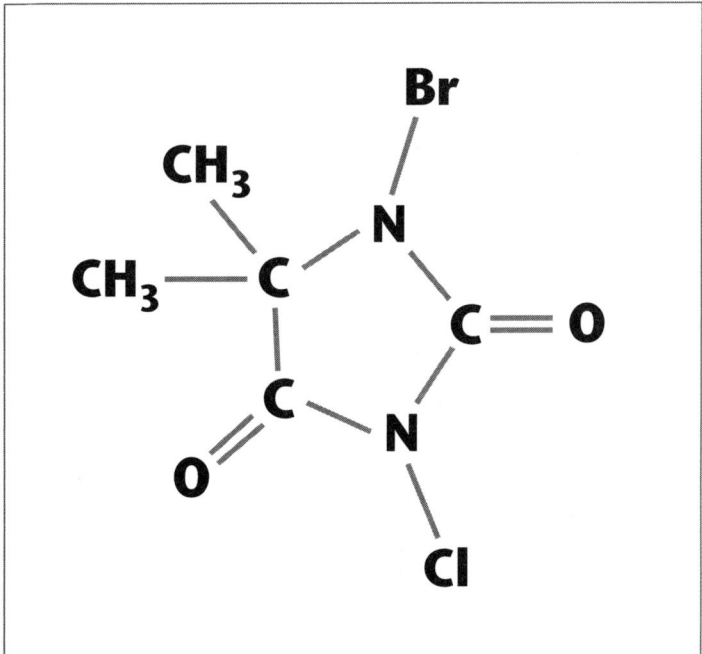

Fig. 7–14 *Bromo-chloro Hydantoin*

Bromine will also form halogenated organics. Although these compounds are believed to be less persistent in the environment than chlorinated organics, they have caused some to view bromine treatment with concern.

Practical Example 7-3

Cooling water in an open recirculating system at a manufacturing plant was placed on an alkaline treatment program, in which the pH typically is near 8.5 or a bit above. Gaseous chlorine had once served as the microbiocide, but its effectiveness is greatly reduced at this pH. The company's water treatment consultant installed a simple pot feeder on a sidestream loop of the cooling tower. Plant operators periodically add bromohydantoin to the pot feeder. The total cooling water flow rate is 6000 gpm. An average flow of 3 gpm through the pot feeder generally maintains a free bromine concentration of 0.2 ppm in the cooling water. Certainly this is not the most sophisticated treatment system, but for a firm that has had little capital for major projects, the system has performed adequately.

Chlorine Dioxide

Chlorine dioxide (ClO_2) offers several advantages. These include the following:

- Powerful oxidizer
- Not pH sensitive like HOCl
- Does not form halogenated organics
- Does not react with ammonia

The primary drawback is that chlorine dioxide must be generated on-site because it is much too unstable to be transported. It is also quite a bit more expensive than chlorine or bromine, and since it does not hydrolyze but remains as a dissolved gas, is readily aerated out solution as it passes through a cooling tower.

Two principal methods of ClO_2 production are presently available, chlorine/hypochlorite generation and acid activation of sodium chlorite.

The main ingredient in both is sodium chlorite (NaClO$_2$), which is generally supplied as a 25 percent solution.

$$2NaClO_2 + Cl_2 \rightarrow 2ClO_2 + 2NaCl \qquad [Eq.\ 7\text{–}19a]$$

$$2NaClO_2 + 2NaOCl + 2HCl \rightarrow 2ClO_2 + 3NaCl + H_2O \qquad [Eq.\ 7\text{–}19b]$$

$$5NaClO_2 + 4HCl \rightarrow 4ClO_2 + 5NaCl + 2H_2O \qquad [Eq.\ 7\text{–}20]$$

Of these production methods, the first two have been the most popular for large cooling systems. However, the latter method offers an advantage because no chlorine or hypochlorite is needed. As is similar with gaseous chlorine, an eductor system is employed to mix the sodium chlorite with the activating agents. The sodium chlorite may be stored in a bulk tank for feed to the system.

Chlorine dioxide residuals similar to those for chlorine (0.1 to 0.5 ppm) are effective in killing microorganisms. In fact, chlorine dioxide may be even more potent than chlorine, in part because it is not affected by pH as is hypochlorous acid.

As with the other oxidizing biocides, safety is very important while using chlorine dioxide. Sodium chlorite is a powerful oxidizer and will react with carbon-based compounds. Sodium chlorite is generally safe in its liquid form, but be sure to clean up any sodium chlorite spills. I was once at a utility in which some sodium chlorite leaked from a supply truck onto the ground. The product dried before maintenance personnel could clean it up. A maintenance technician came out to look at the spill and made the mistake of walking onto the spilled material. The ground erupted in smoke at his feet, as friction from his shoes ignited the unstable material. I had not seen somebody move that fast in a long time.

Chlorine dioxide can be an excellent biocide, but its high cost and tendency to come out of solution in cooling towers restrict its usefulness.

Ozone

Ozone use has increased in recent years, and the technology has matured enough that it is being used for treatment of some smaller, industrial recirculating systems. Ozone (O_3) is a generally short-lived, extremely powerful oxidant that must be generated on-site. It is produced by passing an oxygen or air stream through a high-voltage current. The air stream is then bubbled into the recirculating water.

The primary advantages of ozone are that it is a very powerful oxidizer that does not produce any harmful reactants such as halogenated organics. At this time, ozone is not extensively used for treatment of utility or large industrial cooling towers. In part this is due to the fact that cooling water often has enough organic material to create a high ozone demand, which uses up much of the biocide before it can attack microorganisms. Work is continuing in this field.

Non-Oxidizing Biocides

Due to the safety and environmental concerns regarding oxidizing biocides, non-oxidizing biocides have received more attention. Sometimes the non-oxidizers are used in a joint program with oxidizing reagents to attack microorganisms.

A number of non-oxidizing biocides are currently available. Some of the most popular today include isothiozolone (Figure 7-15), quaternary amines (Figure 7-16), and bromonitropropanediol (BNPD, Figure 7-17). These biocides work by either reacting with the cell wall or interfering with the organism's metabolic processes. Quaternary amines penetrate into cell walls and disrupt transport of products through the cell wall. BNPD and isothiozolone react with internal cell components or interfere with protein making processes.

Microorganisms can develop a resistance to the non-oxidizing biocides, so careful planning may be needed for them to be effective. One method is to use non-oxidizers as a supplement to oxidizing biocides. A periodic batch dosage can be used to shock the microbes and kill those that may have survived the oxidizing chemical. For example, as a supplement to the bromine feed mentioned in Practical Example 7-3, an automatic controller feeds isothiozolone once per week. Another alternative is to feed two non-oxidizers on a semi-continuous basis, where each is alternated periodically (perhaps even daily) so that the microorganisms do not build up any tolerance.

CHAPTER 7: COOLING WATER CHEMISTRY

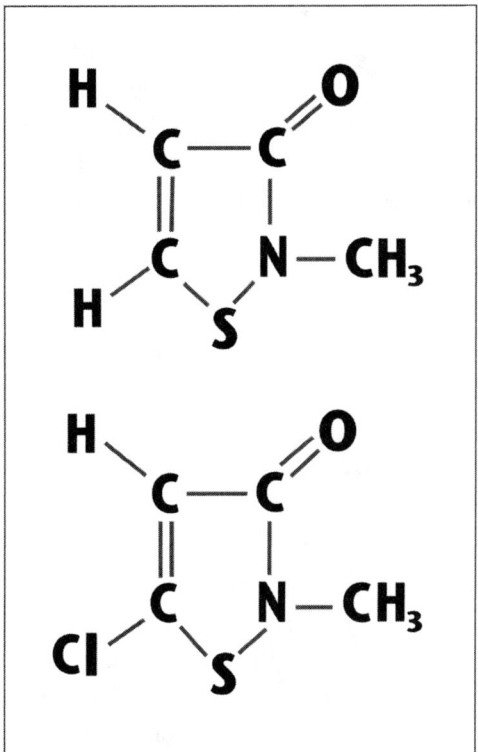

Fig. 7–15 *Isothiozolone Structures*

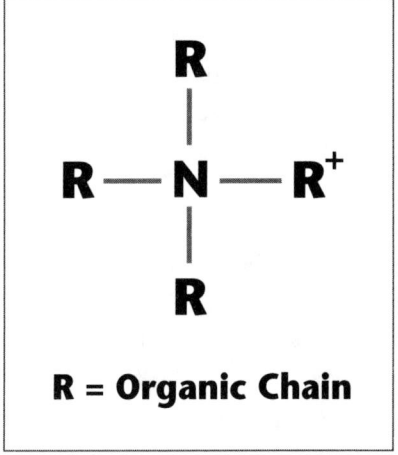

Fig. 7–16 *Quaternary Amines*

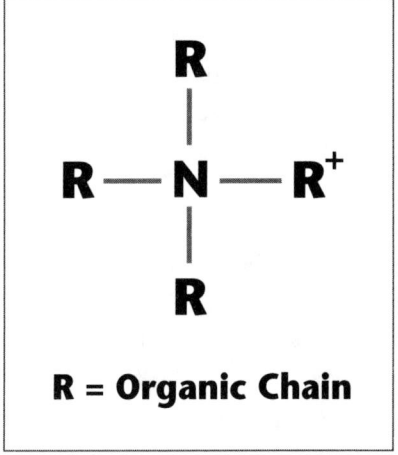

Fig. 7–17 *Bromonitropropanediol (BNPD)*

Just like oxidizing biocides, non-oxidizers can have an effect on the environment if they leave the plant in the cooling discharge. Various treatment methods are possible to neutralize many of these chemicals. For instance, some of the non-oxidizers, isothiozolone and BNPD in particular, will decompose when treated with sulfite. Bentonite clay, if added to the discharge, provides absorption sites for compounds such as the quaternary amines.

Macrofouling

Fouling caused by organisms whose individual members are visible to the naked eye is known as macrofouling. Macrofouling may be caused by a number of different creatures, but the two most troublesome freshwater species by far have been Asiatic clams and now zebra mussels.

Macrofouling by clams occurs when the creature dies and the shell is washed into sensitive flow areas such as a condenser. Asiatic clams have proven troublesome because shell sizes typically range from about $\frac{1}{2}$" to 1" in size. Since condenser and heat exchanger tubes often are of this size, the clam shells make a good fit. A growth in Asiatic clams within a cooling system could play havoc with operation.

Zebra Mussels

These creatures were inadvertently introduced to the Great Lakes in the mid-1980s, most probably from discharge of ballast water of a European freighter. Zebra mussels get their name from their alternately striped shells.

Zebra mussels are freshwater bivalves that may grow to about 5 cm in length. The principal difficulty comes from their colonization patterns. Zebra mussels are microscopic when first spawned, being about 40 microns in size. They have no shell at first, and the young mussels are known as veligers. Within just a few weeks after hatching, however, the veligers develop a shell and begin to search for a place to settle. When the mussel has found a location, it extends fibers, known as byssal threads, that attach permanently to the surface. Zebra mussels are troublesome primarily for four reasons:

- The microscopic larvae (veligers) can be carried a long distance by water currents.
- The veligers are much too small to be stopped by travelling screens at water intakes.
- The mussels prefer flowing water, such as cooling water intakes, where nutrient supplies are plentiful. However, if water flow rates are higher than 3 or 4 feet per second they may not settle.
- They will attach to many different hard surfaces, including each other, to form dense, thick masses.

This latter aspect is what makes zebra mussels particularly annoying. Densities of up to 80,000 creatures per square foot have been found at locations in the Great Lakes. The mussels can clog trash racks, travelling screens, auxiliary cooling water lines, and fire lines. Masses that break loose in the main cooling water line will plug condenser tubes. One of the classic photos of a mussel colony has appeared in several publications. It shows an abandoned automobile pulled from Lake Erie. The car's shape is apparent, but the body is almost entirely covered with mussels. Zebra mussels will even attach to other aquatic organisms such as crayfish.

Various control methods have been investigated for control of these creatures. Some of them include electrification of trash racks, application of fouling or slick coatings to intake pipes, and even use of acoustics. None have yet been adopted on an industrial level. More effective have been mechanical removal, chemical treatment and thermal shock. The first treatment requires a system shutdown and draining so that personnel can physically scrape the mussels off equipment. As you can imagine, this is time consuming, operationally expense, and tedious. It does not take much imagination to figure out what the smell is like in an intake structure filled with dead mussels. Obviously, mechanical removal is not the most preferred treatment method.

Chemical Treatment

Plant personnel can combat mussel infestations by diligently operating the biocide system, even though feed may not be continuous. Infant zebra mussels have no shell, and it takes several days for one to develop. Regular oxidant feed to the cooling water will kill the veligers before the shell fully develops. Destruction of mature mussels takes a more vigorous effort. Adult mussels can sense oxidizing biocides, and will simply close up if they detect a residual oxidant. They can stay closed for at least several days, and can certainly survive an intermittent oxidant feed. One method is to initiate continuous oxidant feed, and detoxify the discharge. Eventually, the mussel will have to open, at which time the oxidant can kill it. Such a treatment can only be used if the plant's environmental permit allows for it. However, the EPA has granted variances for this type of application to facilities threatened by zebra mussels.

An alternative treatment is feed of a non-oxidizing biocide. Mussels cannot readily detect these chemicals and will filter the water unwittingly. Non-oxidizers damage zebra mussel cells just as they do microbiological cells. Quaternary amines have proven to be effective. Again, however, chemical feed must be approved by the EPA or state environmental agency that oversees the plant.

Thermal Shock

Zebra mussels are not very tolerant to warm water (temperatures above 100°F). It is possible to destroy mussels by thermal shock, and this method is feasible at facilities in which warmed cooling water from the condenser can be recirculated to the cooling water intake. A winter-time application is best, when the mussels have become accustomed to cold water. Good results have been obtained when mussels acclimated to 50°F water were suddenly subjected to 100°F water. Since many utilities and industrial plants do not have this capability, thermal shock is not a universal treatment.

Conclusion

Cooling water treatment is often a complex process that requires close operator attention. Environmental restrictions have not made the job any easier. Alkaline cooling water treatments and chlorine-alternative microbiological programs are gaining popularity. Plant personnel must carefully consider any treatment program and take into account such factors as economy, safety, efficiency, and environmental regulations.

Learning Aide 7-1

COOLING TOWER CALCULATIONS

The primary purpose of a cooling tower is to remove heat while minimizing water usage. Heat is transferred by two mechanisms. Most of the heat transfer takes place due to evaporation of a small amount (generally 1% to 3%) of the cooling water. Latent heat is given up in this phase change. The remainder of the cooling is through sensible heat transfer, in which heat is exchanged without a phase change. A simple formula exists for calculating the approximate amount of evaporation in a cooling tower.

$$E = (f \cdot R \cdot \Delta T) / 1000 \qquad [Eq.\ 7\text{–}21]$$

where

E is the evaporation in gpm

R is the recirculation rate in gpm

ΔT is the temperature difference (°F) between the inlet and outlet cooling water.

f is a correction factor for evaporation

1000 is the approximate latent heat of vaporization

The value of f is dependent on climate, where the following values are reasonable approximations.

f average = 0.75
f summer = 0.85
f winter = 0.65

The text discusses cycles of concentration (C), blowdown (BD), and drift (D). These values along with makeup (MU) may be calculated from the following equations.

- $BD + D = E/(C-1)$
- $MU = E + BD + D$

These calculations, although not exact, can still be valuable when examining the water balance of a system.

Learning Aid 7–2

SIZING A SIDESTREAM FILTER

A simple set of equations is available to size sidestream filters. The calculations are based on percent solids reduction and the cooling tower blowdown rate. As an example, consider a recirculating water that contains 20 ppm of suspended solids (S_i). The desirable concentration is 5 ppm (S_f). Blowdown from the tower is 100 gpm.

First calculate the percent solids reduction, where:

%Solids Reduction (SR) = $(S_i - S_f)/S_i$ • 100

%SR = (20 ppm – 5 ppm)/20 ppm • 100 = 75%

Determine the filtration rate by the following equation:

%Filtration Rate (FR) = ((100/(100 – SR)) –1) • blowdown rate

%FR = ((100/(100 – 75)) – 1) • 100 gpm = 300 gpm

Learning Aid 7-3

Specifications for Sodium Hypochlorite

Sodium hypochlorite is an unstable material and it will decompose into oxygen, sodium chloride, and sodium chlorate ($NaClO_3$). The decomposition rate is affected by temperature and by the action of metals, most notably iron and copper. These metals catalyze the decomposition. To minimize breakdown of stored sodium hypochlorite solutions, the materials specification should contain the following:

- Iron < 0.5 mg/l
- Copper < 1.0 mg/l
- pH — 11.0 to 11.2

Temperature can have a dramatic impact on hypochlorite solutions. For example, the half-life of a hypochlorite solution is reported to be 800 days at a temperature of 59°F. At 77°F, the half-life drops to 220 days, and at 140°F, the half-life is only 3 days! Bulk storage tanks of sodium hypochlorite should be kept as cool as possible, by sun-shading, painting them white, or both.

BIBLIOGRAPHY

Books

Benson, S.W., *Chemical Calculations, Third Edition*, John Wiley & Sons, New York, NY, 1971.

BetzDearborn, Inc., *Betz Handbook of Industrial Water Conditioning, 9th Edition*, Trevose, Pennsylvania, 1991. BetzDearborn is a division of Hercules, Inc.

Bueche, F., *Introduction to Physics for Scientists and Engineers*, McGraw-Hill, New York, NY, 1969.

Buecker, B., *Power Plant Water Chemistry, A Practical Guide*, PennWell Publishing, Tulsa, Oklahoma, 1997.

Busch, D.H., Shull, H., and R.T. Conley, *Chemistry*, Allyn and Bacon, Inc., Boston, Massachusetts, 1973.

Byrne, W., *Reverse Osmosis: A Practical Guide for Industrial Users*, Littleton, Colorado, Tall Oaks Publishing, Inc., 1995.

Cantafio, A.R., ed., *Drew Principals of Industrial Water Treatment, Eleventh Edition*, Boonton, New Jersey: Drew Industrial Division, Ashland Chemical Co., 1994.

Dillon, C.P., *Corrosion Control in the Chemical Process Industries, Second Edition*, Houston, Texas. Published for the Materials Technology Institute by NACE International, 1994.

Meller, F.H., *Electrodialysis (ED) & Electrodialysis Reversal Technology*, Watertown, Massachusetts: Ionics, Inc., 1984.

Owens, D.L., *Practical Principles of Ion Exchange Water Treatment*, Littleton, Colorado, Tall Oaks Publishing, Inc., 1995.

Purolite, *Purolite Product Guide*, The Purolite Company, Bala Cynwyd, Pennsylvania.

The American Society of Mechanical Engineers, *Consensus on Operating Practices for the Control of Feedwater and Boiler Water Chemistry in Modern Industrial Boilers*, ASME, New York, NY, 1994.

U.S. Filter, *Water and Waste Treatment Data Book*, 19th Printing, United States Filter Corp., Palm Desert, California, 1996.

Van DeLinder, L.S., ed., *Corrosion Basics: An Introduction*, National Association of Corrosion Engineers, NACE International, Houston, Texas, 1984.

Technical Reports and Conference Proceedings

"Interim Consensus Guidelines on Fossil Plant Cycle Chemistry," Palo Alto, California, Electric Power Research Institute, June 1986. CS-4629.

"Osmonics Crossflow Filtration Workshop," Minnetonka, Minnesota, Osmonics, Inc., 1995.

"Proceedings of the 60th Annual Meeting, International Water Conference," Pittsburgh, Pennsylvania, Engineers' Society of Western Pennsylvania, 1999.

"Proceedings of the 59th Annual Meeting, International Water Conference," Pittsburgh, Pennsylvania, Engineers' Society of Western Pennsylvania, 1998.

"Proceedings of the 57th Annual Meeting, International Water Conference," Pittsburgh, Pennsylvania, Engineers' Society of Western Pennsylvania, 1996.

"Proceedings of the 19th Annual Electric Utility Chemistry Workshop," Champaign, Illinois, University of Illinois, 1999.

"Proceedings of the 15th Annual Electric Utility Chemistry Workshop," Champaign, Illinois, University of Illinois, 1995.

"Proceedings of the Fifth International Conference on Cycle Chemistry in Fossil Plants," Charlotte, North Carolina, Electric Power Research Institute, 1997.

INDEX

A

Acetic acid, 41

Acid feed (scale control), 287–289

Acid-base reactions, 24–25

Acidity/basicity, xxi, 38–41, 114–116, 193–195, 287–289
control, 193
conditioning, 193–195

Activated carbon filtration, 76–77

Addition/elimination reactions, 26

Admiralty metal, 150–151
et passim

Aerator, 90–91

Air in-leakage, 153

Alarms, 116

Alkaline earths, 11–12, 16

Alkaline treatment (scale), 280, 289–292

Alkalis, 10–11, 16–17

All-volatile treatment, 196, 204, 213–215

Alternative pretreatments, 77–78

Alternative programs (phosphate treatment), 211–215
equilibrium phosphate, 211
phosphate treatment, 211
other boiler water treatment programs, 211–215
chelant treatment, 212–213
polymer treatment, 212–213
caustic treatment, 213
all-volatile treatment, 213–215
oxygenated treatment, 215

Aluminum sulfate, 64

Amines, 81, 193–195, 304–305

Aminoethylenephosphonic acid, 290

Ammonia, 150–151, 193–195, 197

Ammonia-oxygen attack, 150–152

Analytical techniques, 240–241

Anode reaction, 185–186

Anodic inhibitor, 277–278

Aquifers, 35, 47

Atmosphere effects, 44–45

Atomic mass, 3–5

Atomic number, 3–5

Atomic structure/size, 2–3

Atoms/molecules, 2–6
    atomic structure/size, 2–3
    atomic number, 3–5
    atomic mass, 3–5
    isotopes, 3–5
    elements, 4–5
    electronics, 6

B

Barium fluoride, 57

Barium sulfate, 55–56

Bicarbonate, 43

Biocides, 60–62, 75, 241, 292–308

Boiler chemistry, 166–176

Boiler deposits, 163–163

Boiler material composition, 177–180

Boiler water, xviii, 199–224, 234–236
    treatment programs, 204–224

Boiler water treatment programs, 204–224
    early boiler water treatment, 204–206
    coordinated/congruent phosphate treatment, 206–209
    phosphate hideout, 209–210
    alternative phosphate treatment programs, 211–215
    example, 215
    heat recovery steam generators, 216
    steam chemistry, 216–220
    condensate return, 220–221
    layup/off-line corrosion protection, 221–223
    chemical feed systems, 223–224

Boilers, xviii, xxi, 157–159, 163–163, 166–180, 199–224, 234–236
    drum-type, 158
    package drum, 158
    field-erected drum, 158

Bond types, 19–21
    ionic, 19–20
    covalent, 20–21
    metal, 21
    coordination chemistry, 21

Bromine, 300–302

Bromonitropropanediol, 304–305

Buffers, 42–43

C

Calcium carbonate equivalency, 50–52, 54

Calcium carbonate scale, 282–285

Calcium fluoride, 282

Calcium hypochlorite, 298

Calcium phosphate, 282, 285–286, 292

Calcium sulfate, 282, 285–286, 292

Carbohydrazide, 189–191

Carbon, 11–12, 180

Carbon dioxide, 43

Carbonate, 43

Carbonic acid, 141–142

Carboxylic acid, 81

Carryover, 166–169

Cathode reaction, 185–186

Cathodic inhibitor, 277–279

Cation bed performance, 83–85

Caustic, 94, 129, 204, 213
  regeneration, 94
  treatment, 213

Caustic regeneration, 94

Caustic treatment, 213

Cellulose acetate/triacetate, 106–107

Charge neutralization, 62, 65

Chelant, 212–213
  treatment, 212–213

Chelant treatment, 212–213

Chemical bonding, 3, 12, 19–21, 36–37
  bond types, 19–21

Chemical control (oxygen), 187–191

Chemical feed systems, 223–224

Chemical properties, 6–19

Chemical reactions, 22–31
  mole concept, 22–23
  standard temperature and
    pressure, 24
  acid-base reactions, 24–25
  addition/elimination reactions, 26
  substitution reactions, 26
  oxidation-reduction reactions, 26–30
  equilibrium reactions, 30–31

Chemical structure, 6–19

Chemical treatment programs, xviii, 185–224, 307–308
  condensate/feedwater treatment, 185–192
  flow-assisted corrosion, 192–195
  oxygenated treatment, 196–199
  chemistry guidelines, 199–203
  boiler water treatment, 204–224

Chemistry fundamentals, xviii, 1–34
  basic concepts, 6–19

Chemistry guidelines, 199–203
  feedwater, 199
  water chemistry limits, 200–203

Chemistry in generators, 164–165

Chloride, 172–173

Chlorinated isocyanurates, 298–300

Chlorine, 125, 296–303
  fouling treatment, 296–303
  application, 297–298

Chlorine alternatives, 300–302

Chlorine dioxide, 302–303

Chlorine donors, 298–300

Chlorine fouling treatment, 296–303
  application, 297–298
  chlorine donors, 298–300

chlorine alternatives, 300–302
chlorine dioxide, 302–303
Chromium, 180
Clarification, 60, 62–70
Clarifier arrangement, 60, 62–70
Cleaning (reverse osmosis), 117–118
Cleaning chemicals, 117–119
Cleanliness monitoring, 259–261
Closed cooling systems, 270
Closed recirculating cooling system, 265
Coagulants, 64–65, 69, 108
Coagulation, 62, 64–65
Co-current/countercurrent systems, 89–93
Combined water treating, 196
Communications, 250–251
Components (reverse osmosis), 113–114
Concentrate pressure, 116
Concentration, 50–52
Condensate discharge, 148–149, 230–231
Condensate pH control, 193
Condensate polisher, 153–154, 231–232
 effluent, 231–232
Condensate return, 140–143, 220–221, 229–230
Condensate subcooling, 183

Condensate/feedwater systems, 140–143, 148–149, 153–156, 183, 185–193, 220–221, 229–232
 treatment, 185–192
Condensate feedwater treatment, 185–192
 dissolved oxygen control, 186–187
 chemical control of oxygen, 187–191
 safe hydrazine systems, 191–192
 oxygen scavenging, 192
Condensation process, 181–183
Condenser backpressure, 254–258
Condenser cleanliness, 259–263
Condenser performance, 252–258
Condenser tube, 143–147, 149–150, 262
 correction factors, 262
Condenser tube correction factors, 262
Conditioning rack (sample), 247–250
Contaminants (drum boiler), 172–174
 copper, 172
 sodium hydroxide, 172
 chloride, 172–173
 sulfate, 172–173
 iron oxides, 173
 silica, 173–174
 sodium phosphates, 173
 organics, 173
Contamination sources/problems, xviii, 133–134, 172–174
Control guidelines, 225–263
 sample point selection, 225–227

INDEX

makeup treatment system, 227–231
example, 231, 251–252, 256–257
condensate polisher effluent, 231–232
deaerator inlet, 232–233
deaerator outlet, 233
feedwater/economizer inlet, 233–234
boiler water, 234–236
saturated steam, 236–237
main steam, 237
reheat steam, 237–238
steam purity, 238–240
analytical techniques, 240–241
instrument maintenance, 241
sample extraction/conditioning, 241–250
instruments and communications, 250–251
condenser performance, 252–257
condenser cleanliness, 259–263

Cooling systems, 265–270
once-through cooling, 266
open recirculating systems, 266–270
closed cooling systems, 270

Cooling tower calculations, 309–310

Cooling water, xviii, 263, 265–314
systems, xviii, 265–281, 292–308
correction factors, 263
chemistry, 265–314

Cooling water chemistry, 265–314
cooling systems, 265–270
corrosion, 270–281
scale, 270–271, 281=292
fouling, 270–271, 292–308

examples, 278–281, 302
cooling tower calculations, 309–310
sidestream filter sizing, 311
sodium hypochlorite, 313–314

Cooling water correction factors, 263

Cooling water systems, xviii, 265–281, 292–308

Coordinated/congruent treatment, 206–209

Coordination chemistry, 21

Copper, 172, 274–275, 280–281
corrosion, 280–281

Corrosion, xxi, 135–139, 149–151, 156–157, 169–170, 192–195, 198, 204–205, 216, 221–223, 270–281
fundamentals, 135–139
fatigue, 138
flow-assisted, 156–157, 192–195, 198, 216
protection, 221–223
cooling water systems, 270–291
control, 271–275
cell, 273
inhibitors, 277–278

Corrosion (cooling water system), 270–281
influence on mechanical factors, 275–276
non-metallic corrosion, 276
corrosion inhibitors, 277–278
examples, 278–281

Corrosion cell, 273

Corrosion control, 271–275

Corrosion fatigue, 138

323

Corrosion fundamentals, 135–139
   general corrosion, 135–136
   pitting, 136–137
   crevice corrosion, 137
   galvanic corrosion, 137
   erosion-corrosion, 137
   stress corrosion cracking, 137–138
   corrosion fatigue, 138
   intergranular corrosion, 138
   dealloying, 138
   exfoliation, 138
   microbiologically-influenced corrosion, 139

Corrosion inhibitors, 277–278

Corrosion protection, 221–223

Covalent bond, 20–21

Cracking (corrosion), 137–138

Crevice corrosion, 137

Crossflow filtration, 77–78, 102–104

Crystal modifiers, 290

D

Deaerator, 155–156, 232–233
   inlet/outlet, 232–233

Dealloying, 138

Degasifiers, 88–89

Demineralization, 87–93

Demineralizer, 87–93, 98–102
   configuration, 87–88
   troubleshooting, 98

Di-/triprotic acids, 41–42

Dechlorohydantoin, 298–299

Disinfecting, 125–126, 295–308
   methods, 125–126

Dissolved hydrogen monitoring, 240

Dissolved ions, 72–75, 146

Dissolved oxygen control, 186–187

Double-pass system, 112

Drum boiler chemistry, 166–176
   mechanical carryover, 166–168
   vaporous carryover, 168–169
   solids introduction, 169
   superheater exfoliation, 169
   steam chemistry, 169–172
   contaminants, 172–174
   example, 174–175
   startups/shutdowns, 175–176

Drum-type boiler, 158, 166–176
   chemistry, 166–176

E

Earth crust elements/compounds, 45–50

Economizer, 158, 233–234

Electrical charge, 1–2

Electrochemical reaction, 27–28

Electrodeionization, 79, 122–123

Electrodialysis/electrodialysis reversal, 79, 120–122

Electron, 3, 6, 8, 16–17, 185–186

Electronegativity, 16–19, 36

Elements, 4–5

Energy transformation, 133–134

Enthalpy (steam), 181–183

Environmental regulation, 280

Equations, 56–57

Equilibrium phosphate treatment, 204, 211

Equilibrium reactions, 20–31

Erosion-corrosion, 137

Ethylenediamine-tetracetic acid, 212

Examples, 69–70, 118–119, 142–143, 147–152, 174–175, 215, 231, 251–252, 256–257, 278–281, 302

Exchange groups, 81–82

Exfoliation, 138, 169

Extra filtration, 77

F

Fatigue, 138

Feed pump pressure, 117

Feedwater, 154–156, 158, 185–193, 199–203, 233–234
  inlet, 158, 233–234
  treatment, 185–192
  pH control, 193
  water quality, 199–203

Feedwater quality, 199–203

Feedwater/economizer inlet, 158, 233–234

Ferric chloride, 65

Ferric sulfate, 64

Field-erected drum boiler, 158

Fill material (cooling tower), 267–269

Filter media, 71, 127
  selection, 127

Filtration, 60, 70–72, 76–78, 102–104, 119–120, 127, 293

Flocculating/flocculants, 65, 67, 69

Flow control/monitoring, 114–116

Flow-assisted corrosion, 156–157, 192–195, 198, 216
  condensate/feedwater pH control, 193

Fossil fuels, 12

Fouling (cooling water system), 131–132, 270–271, 292–308
  index, 131–132
  non-microbiological control methods, 293–294
  inhibitors, 294
  microbiological fouling, 295
  microbiological organism control, 295–296
  chlorine, 296–300
  chlorine alternatives, 300–302
  examples, 302
  chlorine dioxide, 302–303
  ozone, 303–304
  non-oxidizing biocides, 304–306
  macrofouling, 306–308

Fouling index, 131–132

Fouling inhibitors, 294

G

Galvanic corrosion, 137, 285

General corrosion, 135–136

Grab samples, 227, 230, 232, 234–235, 237–238

Gravity force, 1–2

Groundwater, 46–49, 78–79
  pretreatment, 78–79

Gypsum, 54

H

Heat recovery steam generators, 160–161, 164–165, 193, 216
  chemistry, 164–165

Helium, 45

High concentrate pressure, 116

High permeate conductivity, 116

High permeate pressure, 116

High/low influent pH, 116

High-temperature shutoff solenoid, 248–249

Hollow-fiber membrane, 104

Hot-lime softening, 74

Hydrazine, 188–189, 191–192
  systems, 191–192

Hydrochloric acid, 41

Hydrogen, 40, 45

Hydrogen sulfide, 47

Hydrologic cycle, 35

Hydroquinone, 189–191

Hydroxyethylidene diphosphonic acid, 289–291

Hydroxyl ion, 40

I

Impurities (natural water), 44–57
  atmosphere effects, 44–45
  Earth crust elements/compounds, 45–50
  units of measurement, 50–52
  solubility of minerals, 52–54
  solubility product, 55–56
  equations, 56–57

Independent power producers, xvii

Influent pH, 116

Inhibitors, 277–278, 287–292, 294
  corrosion, 277–278
  scale, 287–292
  fouling, 294

Instruments, 241, 250–251
  maintenance, 241

Intergranular corrosion, 138

Ion exchange, 80–81

Ion exchange resins, 80–81

Ion exchangers, 83–87, 95–98
  vessel performance, 95–98

Ionic bond, 19–20
Ionization enthalpy, 9
Iron oxides, 161–164, 173
  deposition, 161–164
Iron-based alloys, 179
Isothiozolone, 304–305
Isotopes, 3–5

J

Jar testing, 69

L

Langelier saturation index, 282–285, 292
Layup/off-line corrosion protection, 221–223
Le Chatelier's principle, 41, 89
Lime treatment, 72–75

M

Macrofouling, 306–308
  zebra mussels, 306–307
  chemical treatment, 307–308
  thermal shock, 308
Magnesium silicate, 286
Main steam, 237

Makeup water, xviii, 59–132, 139–140, 227–231
  treatment system, 227–231
Makeup water treatment, xviii, 59–132, 227–231
  pretreatment, 60–79
  polishing techniques, 79–102
  crossflow filtration, 102–104
  reverse osmosis, 102–119
  other membrane technologies, 119–123
  alternate disinfection methods, 125–126
  filter media selection, 127
  sulfuric acid/caustic specifications, 129
  silt density index, 131–132
  system, 227–231
  condensate return, 229–230
  condensate pump discharge, 230–231
Manganese, 180
Material composition (boilers), 177–180
  operating temperatures, 178
  iron-based alloys, 179
  nickel-based alloys, 179–180
  chromium, 180
  carbon, 180
  molybdenum, 180
  manganese, 180
  silicon, 180
Mechanical carryover, 166–168
Mechanical factors, 275–276
Memgbrane cleaning, 117
Membrane design, 104–106

Membrane material, 106–107

Membrane technologies, 102–123
  crossflow filtration, 102–104
  reverse osmosis, 102–119
  micro-ultrafiltration, 119–120
  electrodialysis/electrodialysis reversal, 120–122
  electrodeionization, 122–123

Metal bond, 21

Methyl ethyl ketoxime, 189–191

Micro-/ultrafiltration, 119–120

Microbiological fouling, 292–293, 295–296

Microbiological organism control, 295–296

Microbiologically-influenced corrosion, 139, 273–274

Microfiltration, 77–79

Mineral salts, 52–54

Mineral solubility, 52–57
  solubility product, 55–56
  equations, 56–57

Mixed-bed exchanger, 87–88
  configuration, 87

Mixed-bed regeneration, 94–95

Mole concept, 22–23

Molybdenum, 180

Monitoring techniques, 225–263
  sample point selection, 225–227
  makeup treatment system, 227–231
  example, 231, 251–252, 256–257
  condensate polisher effluent, 231–232
  deaerator inlet, 232–233
  deaerator outlet, 233
  feedwater/economizer inlet, 233–234
  boiler water, 234–236
  saturated steam, 236–237
  main steam, 237
  reheat steam, 237–238
  steam purity, 238–240
  analytical techniques, 240–241
  instrument maintenance, 241
  sample extraction/conditioning, 241–250
  instruments and communications, 250–251
  condenser performance, 252–257
  condenser cleanliness, 259–263

N

National Association of Corrosion Engineers, 134

Natural elements, 4–5

Natural water chemistry, 35–57
  self-ionization, 38–41
  acidity/basicity/pH, 38–41
  Le Chatelier's principle, 41
  di-/triprotic acids, 41–42
  buffers, 42–43
  impurities, 44–57

Neutral water treatment, 196

Neutron, 3

Nickel Development Institute, 134

Nickel-based alloys, 179–180
Nitrilotriacetic acid, 212
Nitrogen, 13
Non-metallic bonds, 8
Non-metallic corrosion, 276
Non-microbiological control (fouling), 293–294
Non-oxidizing biocides, 304–306

O

Once-through cooling system, 265–266
Once-through steam generation, 166
On-line instruments, 250–252
On-line samples, 227, 230, 232–233, 235–238
Open recirculating cooling systems, 266–270
Operating temperatures, 178
Orbitals, 8–16, 21
Organics, 173
Oxidation-reduction potential, 114–116, 193, 240–241
Oxidation-reduction reactions, 26–30
Oxidizer removal, 75
Oxygen, 13–14, 36, 186–192
    control, 186–192
    scavengers, 188–192
Oxygen control, 186–192

Oxygen scavengers, 188–192
Oxygenated treatment, 156–157, 196–199, 204, 215
Ozone, 303–304

P

Package drum boiler, 158
Packed-bed demineralizers, 98–102
Particle size, 70–72
Periodic table, 6–16
Permeate conductivity, 116
Permeate pressure, 116
pH (acidity/basicity), xxi, 38–41, 114–116, 193–195, 287–289
    control, 193
    conditioning, 193–195
Phosphate hideout, 209–210
Phosphate treatment, 204, 206–214
    coordinated/congruent, 206–209
    phosphate hideout, 209–210
    alternative programs, 211–215
Phosphonates, 289–291
Phosphono-butane-tricarboxylate, 290
Phosphoric acid, 42
Physical concepts, 1–34
    electrical charge, 1–2
    gravity force, 1–2
    atoms/molecules, 2–6
    chemistry basics, 6–19
    periodic table, 6–16

valence, 16–19
electronegativity, 16–19
chemical bonding, 19–21
bond types, 19–21
chemical reactions, 22–31
scientific notation, 33–34

Piping (reverse osmosis), 114

Pitting (corrosion), 136–137, 273–274

Polarity, 37

Polishing techniques, 79–102
  makeup treatment methods, 79
  ion exchange, 80–81
  exchange groups, 81–82
  strong acid cation resins, 82–85
  weak acid cation resins, 85
  strong base anion resins, 85–86
  weak base anion resins, 87
  demineralizer configurations, 87–88
  mixed-bed exchangers, 87–88
  degasifiers, 88–89
  regeneration, 89–92
  co-current/countercurrent systems, 89–93
  strong base anion regeneration, 92
  weak acid/weak base exchangers, 92, 94
  mixed-bed regeneration, 94–95
  ion exchanger vessel performance, 95–98
  demineralizer troubleshooting, 98
  packed-bed demineralizers, 98–102

Polyacrylates, 290–291

Polyamide, 106

Polymaleates, 290

Polymer, 65, 204, 212–213, 290
  treatment, 212–213

Polymer treatment, 212–213

Post-treatment, 75

Practical scale index, 284

Precipitation, 35, 44–45

Pressure filters, 71

Pressure vessels, 109–114

Pretreatment (makeup water), 60–79, 107
  traditional concepts, 60–61
  biocide, 61–62
  clarification, 62–69
  examples, 69–70
  filtration, 70–72
  softening, 72–76
  post-treatment, 75
  activated carbon filtration, 76–77
  extra filtration, 77
  alternatives, 77–78
  groundwaters, 78–79
  reverse osmosis, 107

Priming (steam), 167–168

Proton, 3

Pump pressure, 117

Pumps (reverse osmosis), 113

Q

Quaternary amines, 81, 304–305

R

Recirculating cooling systems, 266–270

Regeneration, 89–95

Reheat steam, 237–238

Reverse osmosis, 11, 102–119
 membrane design, 104–106
 membrane material, 106–107
 pretreatment, 107
 unit design, 109–112
 components, 113–114
 flow control/monitoring, 114–116
 alarms, 116
 high permeate pressure, 116
 high concentrate pressure, 116
 high/low influent pH, 116
 high oxidation reduction potential, 116
 high permeate conductivity, 116
 feed pump pressure, 117
 system size, 117
 cleaning, 117–118
 examples, 118–119

Reverse osmosis unit components, 113–114
 pumps, 113
 pressure vessels, 113–114
 piping, 114

Ryznar stability index, 284

S

Sample conditioning, 247–251

network, 248–251

Sample extraction/conditioning, 241–251

Sample point selection, 225–227

Sample preparation, 241–251

Saturated steam, 236–237

Scale (cooling water systems), 11, 108, 270–271, 281–292
 calcium carbonate, 282–285
 calcium phosphate, 285–286
 calcium sulfate, 285–286
 silica, 286
 magnesium silicate, 286
 control, 287–292
 inhibitors, 287–292

Scale control, 287–292
 acid feed, 287–289
 alkaline treatment, 289–292

Scale inhibitors, 287–292

Scaling potentials, 282–285

Scientific notation, 33–34

Scraping, 293–294

Self-ionization, 38–41

Shutdowns, xvii, 175–176

Sidestream filter sizing, 311

Silica, 173–174, 292, 296
 deposits, 173–174

Silicon, 180

Silt density index, 108, 131–132

Sodium aluminate, 64

Sodium hydroxide, 129, 172

Sodium hypochlorite, 298, 313–314
Sodium phosphates, 173
Softening, 61, 72–76
Solids introduction, 169
Solubility (minerals), 52–54
Solubility products, 55–56
  constants, 55–56
Spiral-wound membrane, 104–105
Stainless steel, 136–137
Standard temperature and pressure, 24
Startups/shutdowns, xvii, 175–176
Steam chemistry, 169–172, 216–220
Steam generating network, xviii, 139–166
  makeup water, 139–140
  condensate return, 140–142
  examples, 142–143, 147–152
  steam surface condenser, 143–147
  water-side corrosion, 149
  steam-side corrosion, 150–151
  steam impingement, 152
  tube sheet leaks, 152
  tube failure due to vibration, 153
  air in-leakage, 153
  condensate polisher, 153–154
  condensate/feedwater system, 154–156
  oxygenated treatment, 156–157
  boilers, 157–159
  heat recovery steam generators, 160–161, 164–165
  iron oxide deposition, 161–164
  chemistry in generators, 164–165
  once-through steam generation, 166
Steam generation fundamentals, 133–183
  contamination sources/problems, 133–134
  corrosion, 135–139
  steam generating network, 139–166
  drum boiler chemistry, 166–176
  boiler material composition, 177–180
  condensation process effect, 181–183
Steam impingement, 152
Steam plant operation, xvii–xviii
Steam purity, 216–220, 238–240
Steam sampling, 143, 236–240, 242–247
Steam separation, 167
Steam solubility, 170–172
Steam surface condenser, 143–147
Steam-side corrosion, 150–151
Steel composition, 177–180
Stress corrosion cracking, 137–138
Strong acid cation resins, 81–85
Strong base anion regeneration, 92
Strong base anion resins, 81, 85–86, 92
  regeneration, 92
Substitution reactions, 26
Sulfate, 172–173

Sulfide, 274
Sulfonic acid, 81
Sulfuric acid, 129
Superheater exfoliation, 169
Surface water, 48–49
Suspended solids, 48, 62–72, 147, 292
System size (reverse osmosis), 117

T

Temperature effect, 54
Terminal temperature difference, 253–254
Thermal shock, 308
Thin-film composites, 106–107
Total dissolved solids, 147, 174–175
Traditional concepts, 60–61
Tube failure, 153
Tube sheet leaks, 152
Turbidity, 68
Twin-bed ion exchange system, 98–102

U

Ultrafiltration, 77–79
Ultraviolet light, 125–126

Unit design (reverse osmosis), 109–112
Units of measurement, 50–52
Upset condition, xvii, xxi, 68, 164

V

Valence, 16–19
Vaporous carryover, 168–169
Vibration, 153

W

Water analysis, 52, 69
Water chemistry, xviii, 35–57, 200–203
   limits, xviii, 200–203
   self ionization, 38–41
   acidity/basicity/pH, 38–41
   Le Chatelier's principle, 41
   di-/triprotic acids, 41–42
   buffers, 42–43
   natural water impurities, 44–57
Water quality, 46–49, 199–203
Water softening, 61, 72–75
Water solubility, 37, 47
Water supply, 48–49
Water treatment, xviii, 59–132, 185–192, 204–224, 227–231
   pretreatment, 60–79
   polishing techniques, 79–102

crossflow filtration, 102–104
reverse osmosis, 102–119
other membrane technologies, 119–123
alternate disinfection methods, 125–126
filter media selection, 127
sulfuric acid/caustic specifications, 129
silt density index, 131–132

Water-side corrosion, 149

Weak acid/weak base exchangers, 92, 94

Weak acid cation resins, 81, 85

Weak base anion resins, 81, 87

Z

Zebra mussels, 316–31